Band 22 · November 1986 **11**

isotopenpraxis

AKADEMIE-VERLAG · BERLIN

ISSN 0021-1915 Isotopenpraxis, Berlin **22** (1986) 11, 377–416

An unsere Autoren

Die Zeitschrift veröffentlicht Beiträge zu folgenden Themen: Kontrolle und Analyse radioaktiver und stabiler Isotope, Anwendung radioaktiver und stabiler Isotope in allen Gebieten der Forschung und Technik, der Industrie, Landwirtschaft und Medizin, Herstellung radioaktiver und stabiler Isotopen sowie von markierten Verbindungen, kernelektronische Geräte für die Isotopenanwendung (Dickenmesser, Dichtemesser u. ä.) Detektoren für isotopentechnische Arbeiten und den Strahlenschutz in Isotopenlaboratorien, Meßtechnik und Strahlenschutz, Dosimetrie in Isotopenlaboratorien, kerntechnische Geräte für die Isotopenanwendung (Beton, Bestrahlungsanlagen u. ä.), Strahlenchemie, Radiochemie, ökonomische und perspektivische Fragen, Fragen der Weiterbildung.

1. Aufnahme von Manuskripten

Zur Veröffentlichung angenommen werden nur druckreife, noch nicht anderweitig publizierte Beiträge in deutscher, englischer oder russischer Sprache zum Themenbereich der Zeitschrift und zwar

- Originalarbeiten
- Kurze Mitteilungen
- Aus der Praxis
- Zusammenfassende Berichte
- Tagungsberichte
- Informationen.

Sind der Beitrag oder Teile daraus bereits veröffentlicht oder ist eine anderweitige Veröffentlichung vorgesehen, so muß dies der Redaktion mit genauer Angabe der Zweitveröffentlichung mitgeteilt werden. Mit der Annahme des Manuskriptes behält sich der Verlag das Recht der Vervielfältigung, Verbreitung und Übersetzung innerhalb der gesetzlichen Schutzfrist vor. Auszugsweiser Nachdruck oder photomechanische u. a. Wiedergabe sind an eine schriftliche Genehmigung des Verlages gebunden. Manuskripte, Zeichnungen und Photos von veröffentlichten Arbeiten werden nur auf den besonderen und bis zum Zeitpunkt der Veröffentlichung mitgeteilten Wunsch des Autors zurückgesandt.

Bei der unverlangten Einreichung von Manuskripten wird empfohlen, eine Befürwortung eines der Herausgeber oder Redaktionsausschußmitglieder beizufügen.

Manuskripte sind einzureichen bei der Redaktion, PSF 370, DDR-7010 Leipzig

2. Form der Manuskripte

Die Autoren werden gebeten:

2 Manuskripte, eines davon in Original-Maschinenschrift, einseitig und mit 2zeiligem Abstand einzureichen; gut leserliche Photokopien sind zulässig; auf dem 1. Blatt (unter Weglassung der Bezeichnung)

den Titel der Arbeit
Vorname-Initialen und Namen sämtlicher Autoren
gegebenenfalls die Institution, in der sie tätig sind
eine kurze Zusammenfassung der Arbeit von höchstens 15 Zeilen
Keywords in alphabetischer Reihenfolge und
als Fußnote die Anschrift der Institution anzugeben;

wenn möglich, Zusammenfassung, Titel und Autoren *englisch* (bei deutschem oder russischen Manuskript) bzw. *deutsch* (bei englischem Manuskript) 2fach beizufügen;

Literaturhinweise (fortlaufend numeriert), Fußnoten, Tabellen, Abbildungen (in 2—3facher Veröffentlichungsgröße), Abbildungsunterschriften (getrennt von den Abbildungen und Zeichenerklärungen, Quellenangaben, falls erforderlich, und gegebenenfalls mit einem Vermerk für besondere Wünsche bezüglich der Abbildungsgröße), private und Dienstanschrift der Verfasser mit akademischen Titeln und ausgeschriebenen Vornamen auf besonderen Blättern ans Ende des Manuskriptes zu stellen;

Wünsche für den Kolumnentitel (Seitenüberschrift) bekanntzugeben;
Absätze durch Einzüge von mindestens 3 Anschlägen kenntlich zu machen;
die Maßeinheiten des SI-Systems („Systeme International") zu verwenden;
Korrekturen im Manuskript unmittelbar an der zu korrigierenden Stelle anzubringen;

Formeln deutlich lesbar und möglichst groß einzutragen;
Farbabbildungen und nachträgliche Änderungen oder Ergänzungen in Abbildungen zu vermeiden;

hervorzuhebende Textstellen zu unterstreichen, und zwar halbfetten Satz mit einer geraden schwarzen Linie, kursiven Satz (Schrägschrift) mit einer blauen Wellenlinie (alle Eigennamen). Sperrungen gestrichelt;
besondere Buchstaben durch farbige Unterstreichung zu kennzeichnen, und zwar Formelzeichen in

Griechisch-Kursiv (α, β, ...) rot,
Kursiv (*A*, *B*, ...) blau
halbfetter Kursiv (**A, B, ...**) 2 Linien blau (Vektoren),
halbfetter Grotesk (**A, B, ...**) 2 Linien violett (Tensoren);

bei Literaturangaben auf Vollständigkeit zu achten. In nachstehender Reihenfolge müssen angegeben sein:

a) bei *Zeitschriften*: Vorname-Initialen und Namen sämtlicher Autoren, Titel der Zeitschrift (nach „Periodica Chimica" oder möglichst wenig gekürzt), Bandnummer (unterstrichen), Jahreszahl (in Klammern), Heftnummer (nur bei Zeitschriften, die jedes Heft mit S. 1 beginnen) sowie die Seitenzahl, auf der die zitierte Arbeit beginnt.

b) bei *Büchern*: Initialen der Vornamen und Name des Autors (Herausgeber werden durch den Zusatz „ed." gekennzeichnet), voll ausgeschriebener Titel des Buches in der Originalsprache, evtl. Bandnummer, voll ausgeschriebener Name des Verlages, Nummer der Auflage (außer der ersten), Erscheinungsort(e), Erscheinungsjahr (hier nicht in Klammern setzen!) und evtl. Kapitelnummer, Seitenzahl u. ä.

c) bei *Sammelbänden, Tagungsberichten*: Es soll zusätzlich der Titel des zitierten Beitrags angegeben werden. Die Seitenzahl muß unbedingt angegeben werden. Aus dem Zitat muß hervorgehen, ob es sich um einen Tagungssammelband oder um einen noch nicht veröffentlichten Vortrag handelt. Zwischen Tagungsort und Erscheinungsjahr des entsprechenden Tagungsberichts ist zu unterscheiden.

3. Korrekturen

Der Autor erhält zwei Umbruchabzüge, von denen einer möglichst schnell korrigiert an die Redaktion zurückgereicht werden soll. Auf Wunsch des Autors kann die Korrektur auch von der Redaktion gelesen werden.

Durch nachträgliche Manuskriptänderungen verursachte Korrekturkosten gehen, sofern sie 10% der Satzkosten übersteigen, zu Lasten des Autors.

Der Autor erhält 50 Sonderdrucke kostenlos. Weitere Sonderdrucke können bis zur Rückgabe der Korrekturabzüge auf Kosten des Autors bei der Redaktion bestellt werden.

Bezugsmöglichkeiten

Bestellungen sind zu richten

- **in der DDR** an den Postzeitungsvertrieb unter Angabe der Kundennummer des Bestellenden oder an den AKADEMIE-VERLAG BERLIN, Leipziger Straße 3—4, PF-Nr. 1233 DDR-1086 Berlin;
- **im sozialistischen Ausland** an eine Buchhandlung für fremdsprachige Literatur oder an den zuständigen Postzeitungsvertrieb;
- **in der BRD und Berlin (West)** an eine Buchhandlung oder an die Auslieferungsstelle KUNST UND WISSEN, Erich Bieber OHG, Wilhelmstraße 4—6, D-7000 Stuttgart 1;
- **in den übrigen westeuropäischen Ländern** an eine Buchhandlung oder an die Auslieferungsstelle KUNST UND WISSEN, Erich Bieber GmbH, General Wille-Str. 4, CH-8002 Zürich;
- **im übrigen Ausland** an den Internationalen Buch- und Zeitschriftenhandel; den Buchexport Volkseigener Außenhandelsbetrieb der Deutschen Demokratischen Republik, Postfach 160, DDR-7010 Leipzig; oder an den AKADEMIE-VERLAG BERLIN, Leipziger Straße 3—4, DDR-1086 Berlin.

Zeitschrift „Isotopenpraxis"

Herausgeber: Zentralinstitut für Isotopen- und Strahlenforschung der Akademie der Wissenschaften der DDR.
Verlag: Akademie-Verlag Berlin, Leipziger Straße 3—4, DDR-1086 Berlin, PF-Nr. 1233; Fernruf: 2236221 oder 2236229; Telex-Nr.: 114420;
Bank: Staatsbank der DDR, Berlin, Kto.-Nr.: 6836-26-20712.
Chefredakteur: Prof. Dr. H. Hübner, Leipzig.
Anschrift der Redaktion der Zeitschrift Isotopenpraxis: Prof. Dr. Hübner, Karl-Marx-Platz, Hauptpostamt, PSF 370, DDR-7010 Leipzig; Fernruf: 281018.
Veröffentlicht unter der Lizenznummer 1527 des Presseamtes beim Vorsitzenden des Ministerrates der Deutschen Demokratischen Republik.
Gesamtherstellung: VEB Druckerei „Thomas Müntzer", DDR-5820 Bad Langensalza.
Erscheinungsweise: Die Zeitschrift „Isotopenpraxis" erscheint monatlich. Die 12 Hefte eines Jahrganges bilden einen Band. Bezugspreis je Band 240,— DM zuzüglich Versandspesen; Preis je Heft 20,— DM. Der gültige Jahresbezugspreis für die DDR ist der Postzeitungsliste zu entnehmen.
Bestellnummer dieses Heftes: 1079/22/11.

AN (EDV) 12831
00900

Isotopenpraxis 22 (1986) 11, S. 377 bis 381

Articles

Trends and Open Questions in Industrial Tracer Applications*)

J. Guizerix

(International Atomic Energy Agency, Industrial Applications and Chemistry Section, Vienna)[1])

The assessment of trends in a given area can be considered as a projection for the future of a present situation resulting from an evolution in the past, analog to a problem of prediction in physics based on the estimation of first, second and third derivative. This exercise is a subjective one when, by analogy, the second and third derivatives are involved whereas main trends related to the first derivative are objectively perceived.

As an introduction we will make a distinction between methodology, technology and nuclear technology and present an outline of tracer methodology in order to define the matter of our subject. Then we will develop:

- *trends in tracer theory;*
- *trends in tracer technology;*
- *trends in tracer applications;*
- *open questions.*

Die Einschätzung von Trends in einem bestimmten Gebiet kann als Voraussage für die Zukunft betrachtet werden, die von der gegenwärtigen Situation und der Entwicklung in der Vergangenheit ausgeht. Dies ist analog zum Voraussageproblem in der Physik, wo erste, zweite und dritte Ableitungen grundlegend sind. Aber genaue empirische Werte für zweite und dritte Ableitungen zu erhalten ist meist schwierig und oftmals subjektiv. Die Haupttrends sind jedoch aus den ersten Ableitungen erkennbar.

Einleitend wird eine Unterscheidung von Methodologie, Technologie und Nukleartechnologie getroffen, um anschließend die Tracermethodologie zu umreißen und dem Thema entsprechend abzugrenzen. Es werden insbesondere entwikkelt und herausgearbeitet:

- *Trends in der Tracertheorie, in der Tracertechnologie und in der Traceranwendung;*
- *einige offene Fragen.*

Keywords.

isotope applications; mathematical models; markov process; matrices; statistical models; stochastic processes; tracer techniques; transfer functions

1. Distinction between Methodology, Technology, Nuclear Technology

The main tools in solving problems regarding tracer applications can be classified as follows:

a) Intuition: individuals possess intuition to a certain degree and the university helps only in shaping this intuition.
b) General methodology: (example: mathematics, physics . . .)
c) Methodology related to the related field: (example: chemical engineering)
d) Tracer methodology.

Tracer methodology implies other methodologies (e.g. hydraulics), tracer theory and tracer technology.

Tracer technology concerns practical or material means to solving a problem; it is the reasons why it seems better, generally speaking, to refer to tracer methodology than to tracer technology which is more restrictive. To win in a game of chess it is necessary to know the rules of the game (technique), but it is not sufficient. Good strategy (methodology) is also required. This distinction is somewhat similar to the difference in the area of computers between hardware and software.

Tracer technology deals with the solution of practical problems such as tracer selection, tracer handling, establishing experimental protocols, data handling and processing, and involves *nuclear technology*, which is more specifically related to the problems of tracer preparation and labelling, problems of detection of nuclear radiation and to safety questions.

2. Outlines of the General Principles of Tracer Applications

It can be considered that all kinds of tracer applications are related to a "system" (a chemical reactor, a sample in autoradiography, a chemical reactor) with an input "flow"

---

*) Presented at the Third Working Meeting on "Radioisotope Applications and Radiation Processing in Industry" in Leipzig, G.D.R. September 23–27, 1985

[1]) Address: Wagramerstrasse 5, P.O. Box 100, A-1400 Vienna, Austria

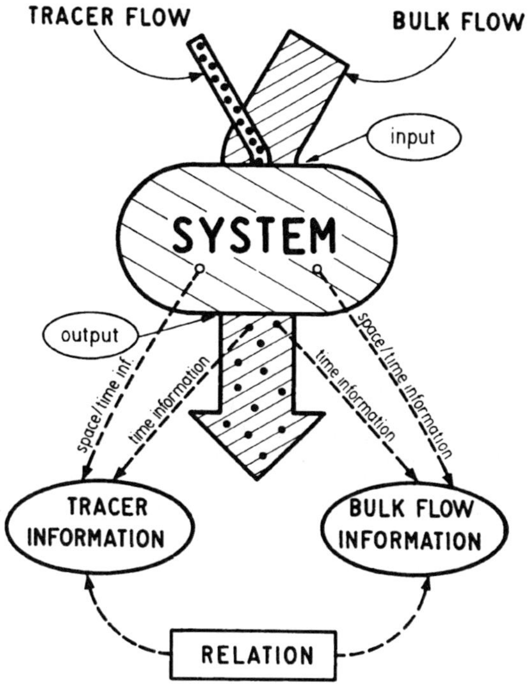

Fig. 1. General principle of tracer applications

and an output "flow" (Fig. 1). This system acts as an operator in mathematics, turning something into something else.

In solving problems we are faced with the requirement for information on the "bulk flow". As we are not able to get this information directly, we add to the bulk flow an auxiliary flow of a substance that can be measured, a "tracer". It is necessary to establish a relation between tracer flow (space/time) information and desired bulk flow information in order to get the latter one.

A classification of tracer information can be shown in Fig. 2.

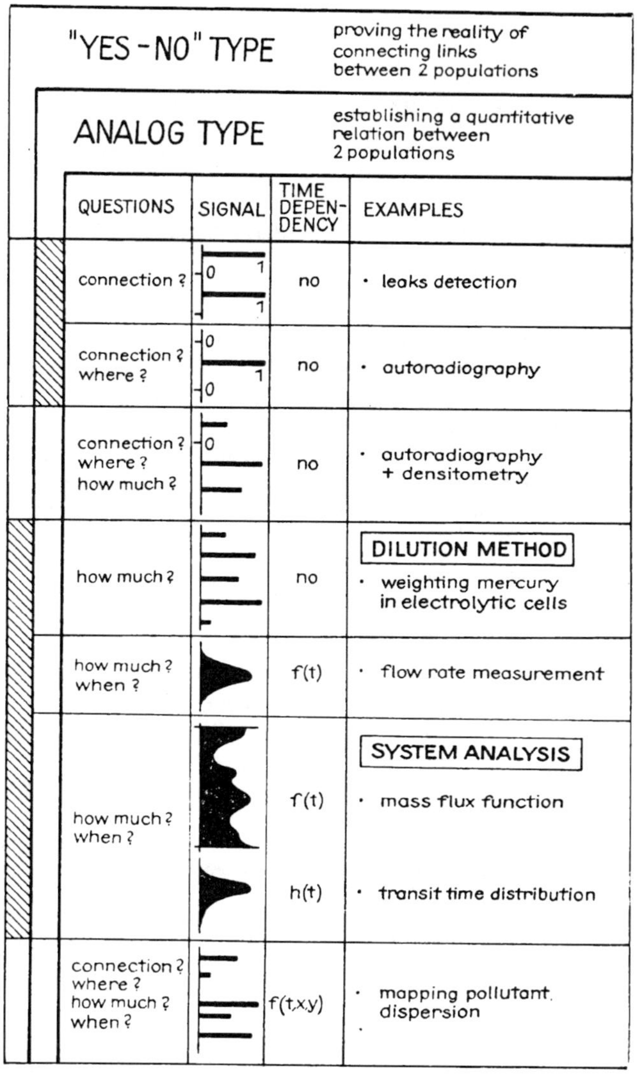

Fig. 2. Nature of information obtained from tracers

3. Trends in Tracer Theory

3.1. *Fundamental facts on tracer theory are ignored or slowly diffused among tracer specialists*

This statement is perhaps surprising but this situation is not abnormal. Making the Fourier series transform of a function is more often realized without knowing the subtleties of the condition to perform this decomposition, as well as it is possible to make a precise measurement of a flow rate with a tracer when the conditions resulting of the theory are only respected and in ignoring standing discussions on the theory.

The levels of comparisons between reality and the analysis of this reality are certainly perceived in different ways by specialists; however, neglecting fundamental facts may result in wrong interpretation or experiments as will be demonstrated by the following three examples.

Example 1: It is incorrect to write a convolution integral with concentration-time functions in the case of a non-stationary system:

$$C(t) \neq \int_{-\infty}^{t} \mathrm{Co}(\theta) \cdot h(t-\theta)\, \mathrm{d}\theta$$

This error is frequent in the field of isotope hydrology. Some development in this matter is given in [1].

Example 2: Ideal injection of a tracer implies perfect mixing of the tracer with the tracee at the entrance to a system and consequently the amount of tracer entering each stream tube has to be weighed by the related velocity of the bulk flow. For example, a punctual injection of a tracer in the feeding flume, 3 × 4 m rectangular, at the entrance of an aeration basin of a sewage work would lead to an incorrect distribution of transit time. One solution consists of injection of the same amount of tracer in many points (40) disposed according to the velocities. The injection of a tracer in a borehole to determine aquifer characteristics has to obey the same conditions. The weighting of the tracer by the natural water velocity is realized if, for example, the borehole is changed into a "good mixer" by a proper mixing technique.

Example 3: The effect of the detection volume in determining distribution of transit time.

Input and output functions to be used in system analysis are mass flux given by the concentration time function multiplied by the flow rate [ref. example 1]. Following the theory these concentration-time functions are "point functions" and consequently a problem arises owing to the fact that in practice the measurement of a concentration always involves a volume. This problem is of importance when the detection volume cannot be considered small against the volume of the system.

Such examples could be multiplied.

From another viewpoint, it is a pity that in the list of conditions to whom a good tracer has to obey, the condition

$$\frac{\text{number of molecules of tracer}}{\text{number of molecules of tracee}} = \text{constant},$$

in all elemental volumes involved in the space of tracer migration. It is, however, the fundamental relation to get information in the bulk flow from tracer information (Fig. 1).

3.2. *Real progress is noted in the theory of constant volume non-stationary systems whereas efforts are observed in the development of open non-stationary systems*

A. Niemi [2] developed a well received analysis of constant volume non-stationary systems which represents a quite significant step towards the development of tracer theory (if the constancy of "flow patterns" is accepted).

Other significant contributions have been presented in the field of open non-stationary systems [3, 4]; however, it appears that further chapters have to be written to reach a better understanding of postulates, better proof of validity of hypotheses, and perhaps better fitting between concepts and reality.

4. Trends in Tracer Technology

4.1. *Slight progress in nuclear techniques due to a saturation effect*

Progress can be observed in last decays for detectors in γ-spectroscopy, some new labelled compounds have been proposed and new isotope generators appeared on the market mainly issuing from medical isotope applications. The interest of these new tools cannot be denied and contributed effectively to the tracer application developments. However, this progress is slight in comparison with the development realized in the past in the field of tracer technology due to the efforts of researchers in the early stages of tracer developments. To create materials tools often precede the stage of ideas to solve problems.

4.2. *Revolutionary step in technological development due to computers*

This step is still under development and new steps are unpredictable. Microprocessors have undoubtedly contributed to the development of tracer applications in facilitating and extending the way to get information.

Example 1: Data aquisition centers. 30 detectors can be used in one experiment.

Example 2: Automation and data processing on the spot. The determination of the flow rate in pipes is possible with an automatic device which proceeds to the injection and the measurement of the tracer, handles the data, introduces corrections and gives the figure of the flow rate.

Example 3: Systematic use of microprocessor (or computer) in automatic data processing.

Some operation as regression analysis and some figures as correlation coefficients, mean value, modal value, standard deviation etc., are systematically and automatically furnished. You push a button and you get the information. That is new, that is interesting, but we have to take care in using this kind of automation and remember the real meaning of such information. That means that we perhaps are not adults in using computers.

5. Trends in Tracer Applications

5.1. *Tracer applications benefit by Cartesian's revenge due to computer advent: solved problems are more complex and of a more interdisciplinary nature*

Tracers are used for building and calibration of more and more complex conceptual models which are represented by "mathematical models". As an example, models for prediction of pollutant transport involve different disciplines such as hydrology, ecology, physico-chemistry and chemical engineering. Computers opened a new area in solving complex problems, making it possible to solve any kind of differential equation. Again we are not yet very advanced in using this new tool, and we must try to avoid errors. For example, confusion generally exists in mathematical models which are more often able to fit data than for prediction.

5.2. *Tracer applications benefit (through tracer methodology) by interdisciplinary linking*

During the last war, system analysis using transfer functions was developed in the field of electronics. Its transposition in the field of mass transfers is pursued with the most beneficial effect. This fusion could be observed in the past when specialists of tracers in hydrology and hydraulics discovered what was carried out in the field of physiology [5] or in chemical engineering [6]. For example, the use of a transit time distribution as an analogy criterion in pilot/real size extrapolation of suspended materials hydraulical systems shows that a link is established between the mass transfer area and hydraulics and we remember that cybernetics was born from the meeting between physiology, mechanics and electronics.

5.3. *Information on tracer applications is being slowly diffused among potential users*

The development of a new methodology is always a time consuming phenomenon; a "time constant" of 15 – 20 years was observed for a penetration of tracer applications among potential users. Some particular features such as the radioactive nature of tracers or such as restrictive regulations on the use of radioactive tracers have in some countries contributed to hindering development.

From another viewpoint, it can be thought that the commercial success of private companies providing tracer services is a significant criterion to assess the usefulness of tracer applications. Such companies, which are faced with the law of supply and demand, will probably contribute in paving the way to further development of tracer applications.

5.4. *A pending question: what is the best? The tracer specialist becomes a specialist in the area of application, or the specialist in the area of application becomes a tracer specialist*

This matter deals with problems of curriculum, problems of optimization. Each of us reaches his own limits rapidly, and cannot have sound expertise in many disciplines. The only solution seems to be the concept of teamwork, research teams, experts teams, etc.

No tendency can be noted in any direction to answer the question asked in the title of this paragraph.

6. Some open Questions

Under "open questions" we understand some methods which seem particularly relevant to tracer applications and which are surprisingly not used or very little used. However, they seem to have the potential to contribute to the further development of tracer methodology. Among these three were selected:

6.1. *Process identification by statistical method (correlation)*

If $u(t)$, $y(t)$, $h(t)$ respectively are the input functions, the output function and the pulse response of a linear system, a convolution integral can be written with the inter-correlation function, $G_{uy}(\tau)$ between $u(t)$ and $y(t)$, the self-correlation function $G_{uu}(\tau)$ of $u(t)$ and $h(t)$

$$G_{uy}(\tau) = G_{uu}(\tau) * h(t) .$$

The symbol $*$ represents a convolution integral [7].
Fig. 3 illustrates two applications of this interesting relation. The first one (Fig. 3a) deals with the determination of pulse response of ore in a problem of hydrotransport [8]. Functions $u(t)$ and $y(t)$ are the signals delivered by a radioactive density gauge disposed in two sections of the pipe. $G_{uu}(t)$ approximates very well a Dirac impulse and the pulse response is directly obtained by calculating $G_{uy}(\tau)$

$$G_{uy}(\tau) \# h(t) .$$

This method can be considered as a generalization of the tracer concept. Information is obtained from the natural variation of a physical characteristic of the flow. The second application is illustrated in Fig. 3b and concerns a "white noise" tracer investigation. In this method, a time-series injection of a small amount of tracers is performed instead of injecting all the tracer in a short time as is normally done. Such a method makes it possible to get a higher precision in determining the transit time distribution with a lower irradiation risk for humans.

6.2. *Markow process*

The Markow process approach is particularly well suited to describe the dispersion phenomenon as will be shown in the following example:

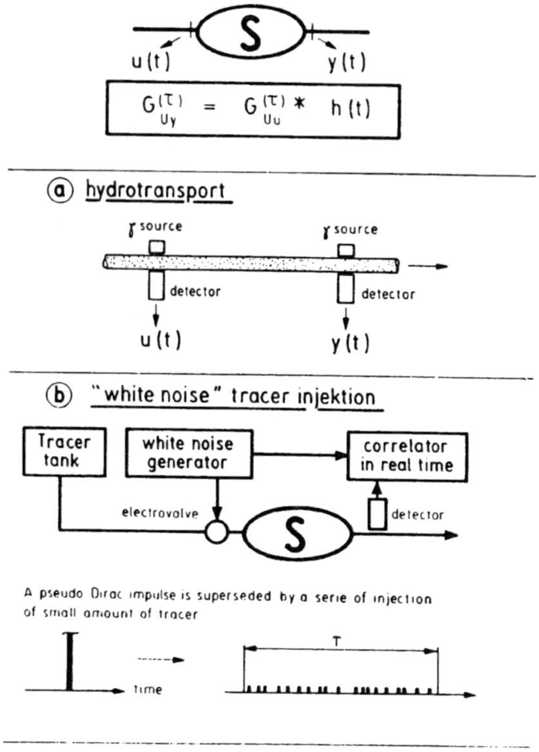

Fig. 3. Process identification by the correlations method

It is supposed that a tracer is injected in a point P of a Section Σ_0 of a river. We want to know the probability for a tracer to appear in different "points" $A, B, C, \dots$ of a down-stream section Σ_m (Fig. 4).

Let $\vec{V}_m$ and $\vec{V}_0$ vectors whose components are probabilities for the tracer to appear in Σ_0 and Σ_m, $||M||$ the transfer matrix of a sub-system defined by section Σ_{k-1}, Σ_k and α the number of sub-system between Σ_0 and Σ_m. Assuming that the transport properties of all sub-systems are identical (first order *Markow* process), then:

$$\vec{V}_m = \vec{V}_0 \, ||M||^\alpha$$

It can be demonstrated that $||M||^\alpha$ converges rapidly when α increases to a matrix all of whose lines are identical due to "mixing" of data. For example, it was possible, by using this interesting property, to understand why the pulse

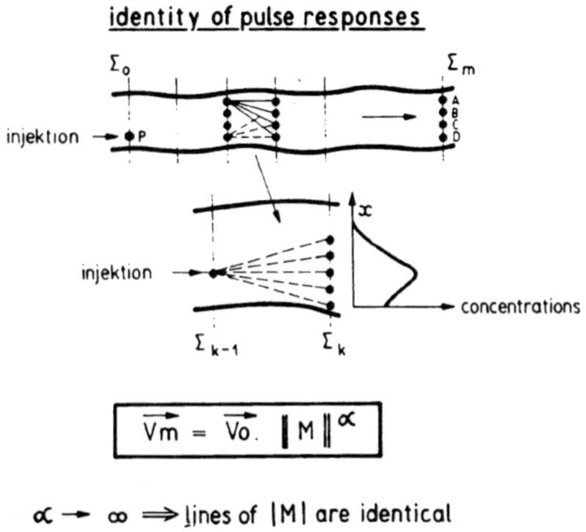

Fig. 4. Markow process approach

responses of a tracer between P and points A, B, and C (Fig. 4) are identical if the distance between sections Σ_0 and Σ_m is great enough [9].

6.3. *Non-linear system approach by functional series representation*

Among other authors, *Amorocho* [10] proposed to link the input and output functions of non-linear systems by a functional series representation:

$$y(t) = \int_{-\infty}^{t} h_1(t, s)\, x(s)\, ds + \int_{-\infty}^{t} \int_{-\infty}^{t} h_2(t, S_1, S_2)\, x(S_1)\, x(S_2)\, dS_1\, dS_2 + \int_{-\infty}^{t} \int_{-\infty}^{t} \int_{-\infty}^{t} h_3(t, S_1, S_2, S_3)\, x(S_1)\, x(S_2)\, x(S_3) \times dS_1\, dS_2\, dS_3) + \ldots$$

This solution, which at one time seemed promising, has not been widely used, probably owing to the fact that a difficult problem of multi-linear deconvolution is implicated. Further development in the direction of a simplification could perhaps contribute to better understanding and greater acceptance of non-linear open systems.

Accepted in revised form January 2, 1986

References

[1] *R. Margrita* et al., "Reflexion sur la théorie des traceurs: Applications en hydrologie isotopique", Isotope Hydrology 1983, IAEA (Proc. Symp. Vienna 1983, 653 – 678)
[2] *A. Niemi*, Internat. J. Appl. Radiation Isotopes 27 (1977) 855
[3] *A. Nir* et al., Tellus 27 (1975) 373
[4] *A. Zuber*, On the interpretation of tracer data in variable flow systems (submitted to Journal of Hydrology)
[5] *C. W. Sheppard*, "Basic principles of the tracer method". New York: John Wiley 1962
[6] *O. Levenspiel*, "Chemical Reaction Engineering". London: John Wiley 1962
[7] *J. Guizerix* et al., "Methodologie d'étude par traceur des transferts de masses", La Houville Blanche 3/4 1976, 187 – 196
[8] *H. Santos-Cottin*, "Application d'une méthode de corrélation à l'identification de processus dans les écoulements diphasiques solides-liquides en conduite", These Université Paul Sabatier Toulouse 1976
[9] *J. Guizerix* et al., "Analyse des informations fournies par les traceurs naturels ou artificiels dans l'étude des systèmes aquifèrs en hydrogéologie", Isotope Techniques in Groundwater Hydrology 1974, IAEA. (Proc. Symp. Vienna 1974)
[10] *J. Amorocho* et al., "Determination of non-linear functional response functions in rainfall – runoff processes", Water Resour. Res. 7 (1971) 1087

Isotopenpraxis 22 (1986) 11, S. 381 bis 384

Ceric-cerous Sulphate and TLD Techniques for Radiation Processing Control in Egypt

A. M. Eid, F. I. A. Said[1])

Radiation Dosimetry Dept., National Centre for Radiation and Technology

S. M. Kamal

Faculty of Science, El-Azhar University, Cairo, Egypt

A review is given of the fundamental dosimetric properties of Ceric-cerous sulphate and natural quartz TL dosimeters in the range 1 – 100 kGy. The development of Ceric-cerous technique is traced from 1980 to the present day with reference to its use as processing control dosimeter. To the same purpose the residual TL of the 280 °C peak of natural quartz is suggested for the first time to extend its gamma-ray response up to 100 kGy. The factors affecting on measurement accuracy for both systems are improved to satisfy the requirements for quality control of irradiated products.

Es wird ein Überblick über die wesentlichen dosimetrischen Eigenschaften von Ce(IV)/Ce(III)-Sulfaten und von natürlichen Quarz-Thermolumineszenz-Dosimetern im Bereich von 1 ... 100 kGy gegeben. Die Entwicklung der Ce(IV)/Ce(III)-Technik kann bis zum Jahr 1980 auf ihre Eignung als Prozeßkontroll-Dosimeter zurückverfolgt werden. Für den gleichen Zweck wird die Rest-Thermolumineszenz beim 280 °C-Peak des Natur-Quarzes für eine Bereichserweiterung bis zu 100 kGy zum erstenmal vorgeschlagen. Die Einflußfaktoren auf die Meßgenauigkeit beider Systeme sind verbessert worden und sie entspricht den Forderungen einer Qualitätskontrolle für bestrahlte Produkte.

Keywords

cerium sulfates; dosimetry; quartz; radiation doses; thermoluminescence dosimeters

1. Introduction

Applying ionizing radiation in industry is nowadays of growing concern to both industrialized and developing countries. Radiation dosimetry provides quality control of all radiation processing, such as sterilization of medical products, preservation of foods and radiation treatment of variety of plastics widely used in modern society.

Many of these processes affect the health and safety of millions of people who use the radiation treated products

[1]) Address: P.O. Box 29, Nasr City, Cairo, Egypt

and it is essential that products are irradiated in calibrated field of radiation or to make accurate radiation dosimetry measurements during the routine operation of the commercial radiation facility. For radiation processing industry, the different dose ranges have been defined.

a) The low dose range (10 Gy – 3 kGy).
b) Mediam dose range (0·1 kGy – 10 kGy).
c) High dose range (5 kGy – 100 kGy).

The main aim of the present investigation is to develop an accurate dosimetry system which can be used during the routine operation of the commercial radiation facility as well as can be mailed world wide to the operators of radiation processing facilities for high level dose standardization and intercomparison on the national and international scale.

For processing dosimetry measurements in Egypt the red perspex technique is used. As this system is dose rate dependent there will be some difference between the calibration curve obtained at constant dose rate and the curve for the varing dose rates in the gamma sterilization plant [1].

In the present study the TLD technique and a modified form of the ceric sulphate dosimetry system are examined in order to estimate the gamma-ray absorbed dose around processing unit. The second aim is to use these techniques for national and international intercomparison.

2. Detectors

The detectors used in the present experimental investigation are ceric sulphate dosimeters and natural quartz TL materials.

2.1. *Cerium Sulphate Dosimeter*

The dosimeter is prepared from 0·1 M $Ce(SO_4)_2$ stock solution: 20·2165 ± 0·001 g $Ce(SO_4)_2$ 4 H_2O (molecular mass: 404·33) was accurately weighed and dissolved in about 250 cm^3 0·4 M H_2SO_4 and completed to 500 cm^3 by 0·4 M H_2SO_4.

Sulphuric acid solution 0·1 M was prepared by using different purity of water; 22·2 cm^3 H_2SO_4 (with a density of $1{\cdot}84 \cdot 10^3$ kg m^{-3}) was mixed in about 500 cm^3 water and completed up to 1000 cm^3.

2.2. *TL Material*

The TL material used during this investigation is natural occuring quartz 99% SiO_2. The material from geological site was grounded and seaved. The grain size between 80 – 170 mesh were retained from TL measurement after washing with distilled water. The TL response of quartz to gamma-ray doses over the range from 100 mGy up to 1 kGy is experimentally known [2].

3. Experimental

3.1. *Calibration Facilities*

For calibration of the used dosimeters, two substandard sources are used: ^{60}Co calibrated unit whose dose rate during these experiments was 718·4 Gy/h and the ^{60}Co irradiation facility at the Egyptian National Centre for Radiation Research and Technology having an activity of $1{\cdot}36 \cdot 10^5$ Ci at the time of the present measurement.

3.2. *Read out Units*

For spectrophotometric measurement, The Pye Unicam Sp 8-200 spectrophotometer is used for measurement of ceric and cerous ion concentration at their absorption maximum 320 and 254 nm [3, 4].

For TLD measurements, the Harshaw 2000 (A + B) system is used where the linear heating rate is 4 °C s^{-1}, the heating cycle is 90 s and the maximum tray temperature is 350 °C. Generally, the samples are annealed before and after irradiation for 1 h at 400 °C using an electric furnace having the temperature range from 50 °C up to 1200 °C with good accuracy (±2%).

4. Experimental Results and Discussion

4.1. *Fundamental Properties of Ceric/Cerous Dosimeter*

The most serious disadvantage of the ceric sulphate dosimeter is its extreme sensitivity to impurities [5]. During this investigation, the effect of impurities on the radiation response linearities were examined. Fig. 1 depict the ceric sulphate dosimeter responses to gamma-rays for sample prepared with different purity of water. From this figure the results indicated that solutions made with monodistilled

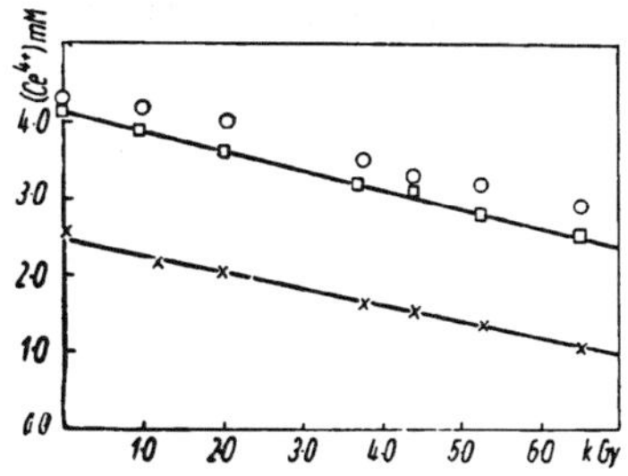

Fig. 1. The effect of impurities on the linearity of ceric and/or cerous ion reduction vs. dose, in 0·4 M H_2SO_4 with different purity water
○ – ceric solution with monodistilled water
□ – ceric solution with bidistilled water
× – ceric/cerous solution with bidistilled water

and mains water do not grow linear with γ-ray dose until certain concentration of cerous ions which corresponding to about 5 kGy the linearity starts. The gamma-ray responses for dosimeters prepared with bidistilled water are linear within ±4%. Hence addition of cerous ions by pre-irradiation improve the linearity as seen in Fig. 1c with high accuracy ±1·5%.

Formation of Ce^{3+} and irradiation temperature. A processing dosimeter has to operate under moderate climate conditions as well as in tropical countries. In this experiment the dosimeters were irradiated to the same dose at different temperatures (5 – 45 °C) adjusted by thermostated water. The results are given in Fig. 2. From this figure no change has been noticed for irradiation temperature between 5 to 45 °C above 45 °C a slight

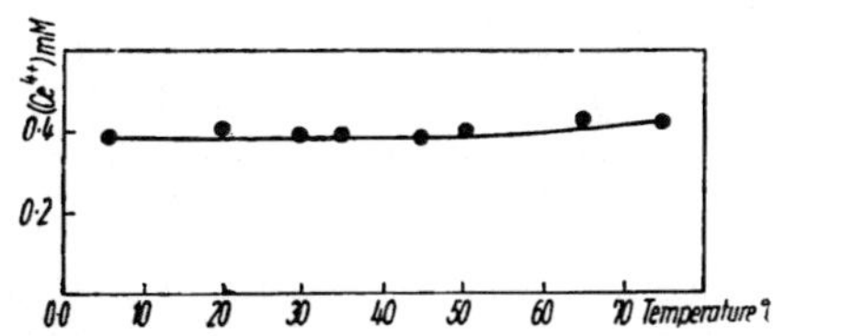

Fig. 2. Temperature effect on the reduction of ceric ions during irradiation

decrease in ceric ion reduction is noticed. This results are in agreement with that of *Hardwick* [6].

From the above results the evaluation of absorbed dose by cerium dosimeter using the experimental factors of the reduction yield [$G(X)$] and molar extinction coefficient (E) of the products may given from the relation:

$$D_{\mathrm{Gy}} = NAF/IG(X)\,EL\,.$$

Where

- N Avogadro's No.
- A absorbance
- I density of dosimeter
- L path length of cell and
- F is the dosimeter conversion factor equal $1{\cdot}6 \cdot 10^{-19}$ Gy/100 eV.

Four routine use, a simple empirical relation is established

$$D_{\mathrm{Gy}} = 824\,Ad.$$

Where d is dilution factor.

4.2. Fundamental Properties of Thermoluminescence Dosimeter (TLD)

Most TLD's are used for gamma-ray dose estimation over the range from few milligrays up to about 5 kGy. In this investigation we tried to extend the range of dosimetry to cover the range of medical products sterilization ($\approx$ 30 kGy) using the residual TL in quartz dosimeter. This residual is a result of the unemitted high temperature glow peaks which created by heavy irradiation. Fig. 3 shows the glow

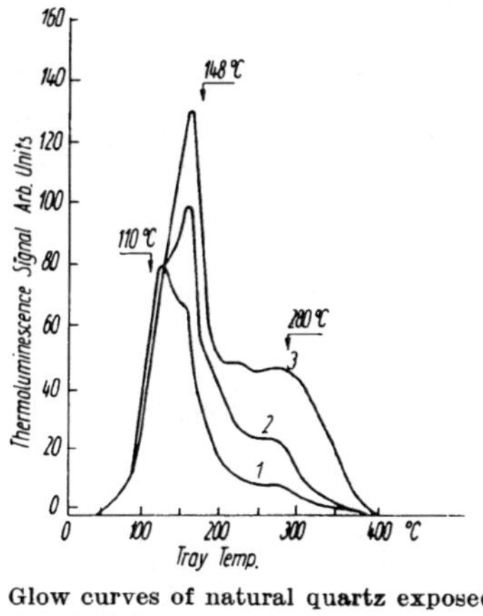

Fig. 3. Glow curves of natural quartz exposed to different gamma-ray absorbed doses
1 – Exposed to 10 kGy
2 – Exposed to 50 kGy
3 – Exposed to 100 kGy

curve structure dependence on the absorbed dose. Below 0·1 kGy our previous studies [2] indicated that the main dosimetric peak is found at 120 °C and a small peak at 280 °C. Above 0·1 kGy new peak appeared at 148 °C which inhances with γ-ray dose and also the 280 °C peak. It is noticed also that the ratio of the 280 °C peak height to the 148 °C peak height increases with γ-ray absorbed rose. This effect may be explained by assuming that there is a competition between different types of traps for capturing the liberated change carriers during irradiation and some traps may be completely filled before the others.

Cleaning of the TL signal stored in the 148 °C for one hour, the residual TL peak height of the 280 °C was found to be increases as the gamma-ray dose is increased above 1 kGy. Fig. 4 shows the variation of TL-signal as a function of ^{60}Co, gammma-ray absorbed dose. From this figure

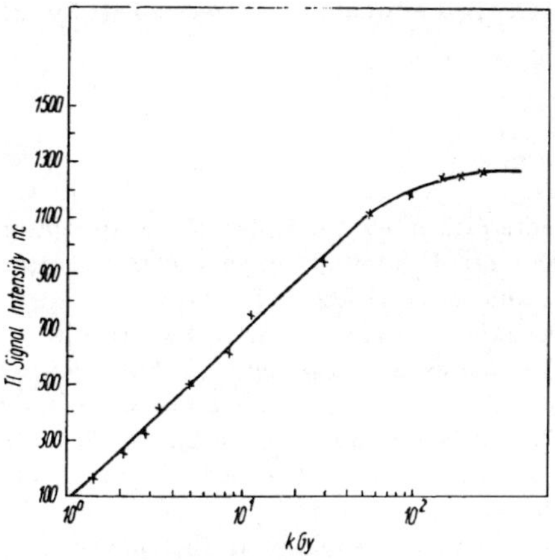

Fig. 4. Variation of residual TL of the 280 °C of natural quartz with gamma-ray absorbed dose

one can deduce that the TL-light intensity of the 280 °C peak increases linearly with the logarithm of the gamma-ray absorbed doses over the range from 1 kGy up to 100 kGy with an accuracy of $\pm 4\%$. Hence the evaluation of unknown dose (D) is straight forward using TL signal and referring it to that of dosimeters which have been exposed to calibration dose (1 kGy – 100 kGy) by the following relation.

$$D_{\mathrm{Gy}} = \mathrm{Exp.}\,(3{\cdot}88 \cdot 10^{-3}\,TL + 6{\cdot}56)\,.$$

Where TL is the thermoluminescence in nanocoloumb (nC) measured under the calibration condition.

Energy Dependence. In case of using radioactive irradiators or X-ray machine of different effective energies. This requires a dosimeter with a response preferably in-

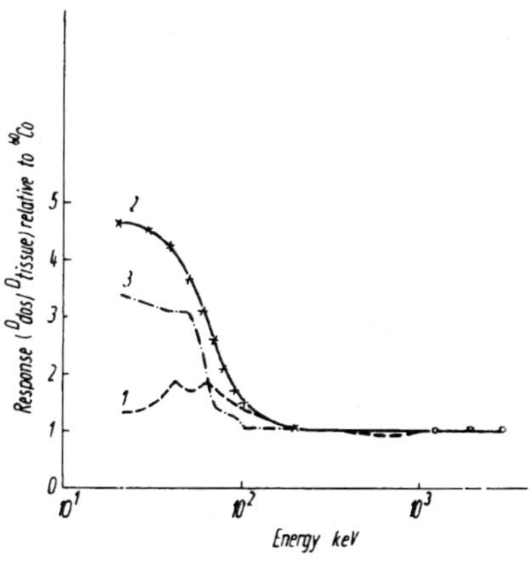

Fig. 5. Variation of response with photon energy
1 – Ceric/Cerous dosimeter
2 – Quartz dosimeter
3 – Their ratio

dependent of the photon energy. For dosimeters under investigation, the energy responses were calculated bases on the cavity theory using the mass energy absorption coefficient of the dosimeter and water [7].

The results are given in Fig. 5, normalized to unity for ^{60}Co-gamma-ray source. From this figure it is clear that ceric/cerous and TLD dosimeters are energy independent above 200 keV (industrial irradiation are usually louded with ^{60}Co or ^{137}Cs gamma ray sources).

Below 200 keV the dosimeters are energy dependent with different ratios. Using the dual system may provide information on beam quality as described by *Rossiter* and *Wood* [8].

5. Precision

Carefully selected dosimetry samples of cerium sulfate and TL of quartz were irradiated under electron equilibrium conditions to medium (1 kGy—10 kGy) and high (5 to 100 kGy) dose ranges of gamma-rays. The irradiation were carried out by using the JS-6500 type ^{60}Co gamma-irradiation plant of Egypt inside the standard sterilization product cardboard box to 30 kGy and 32 kGy (sterilization doses). Free air irradiation were carried out at strategic location in the auxiliary research loop of the above plant. The results are presented in Tab. 1. For the sake of comparison, the results of Red Perspex type 4043E are also included as routine dosimeter.

Tab. 1. Precision of Dosimetric Systems for Measuring Medium and high Doses in Industrial Radiation Processing

Nominal Dose	Measured Dose and Precision Limits		
kGy	TLD	Ceric/Cerous	Red Perspex
30 (P.B.)*	30·7 ± 4·5%	30·8 ± 3%	32 ± 3·5%
32 (P.B.)	32·8 ± 5%	33·6 ± 4%	34·2 ± 2·5%
92·6 (F. air)+	29·1 ± 6·5%	28·6 ± 3·4%	—
19 (F. air)+	18·2 ± 3·7%	18·2 ± 4%	19·4 ± 4%
4·93 (F. air)+	4·91 ± 6%	4·55 ± 3·3%	4·8 ± 6%
2·82 (F. air)+	2·54 ± 7%	—	—

* Irradiation at the top of the Product box (P.B.)
+ Irradiation in Free air (F. air)

From the table, one can deduce that the measured doses in case of irradiation on the top of product box (Cardboard box) are higher than the nominal dose by about 2% for TLD, about 4% for ceric/cerous and about 7% for red perspex. This results may be attributed to different responses of the dosimeters to electrons resulted from gamma-ray interaction with the surrounding materials [9].

Using the derived empirical relations for ceric/cerous and TLD dosimeters showed that both systems with accuracy ±4% for ceric/cerous and 7% for TLD of quartz fulfills the requirements both for research and technological irradiation processes as routine evaluation method (The recommended accuracy value ±5—10%) [10]. Using the TLD calibration curve, an accuracy ±4 % can be achieved with TL of natural quartz.

6. Conclusions

The experimental results indicate that the dosimeters under investigation are capable for monitoring the absorbed dose administered to the goods being sterilized in the Egyptian plant over the range 1—100 kGy with requirements accuracy. Their sensitivities are temperature independent from 20 to 50 °C. The energy dependence of responses for both systems is not appreciable above 200 keV according the *Bragg-Gray* cavity theory. The ratio between the two responses below 200 keV may throw some lights on the mean effective energy of the incident radiation. From the present results it is clear that introducing Ce^{3+} to the dosimeter solution by pre-dose improves its accuracy which show little scatter (3—4%). The dose response curve for TL of natural quartz showed the possibility to use the residual TL of the 280 °C peak of natural quartz in broad dose range up to 100 kGy using the cleaning method of low trapping centres. Having in mind that commercial TLD's were used over the medium dose range up to 5 kGy [11]. The technique is simple, cheap and accurate.

Acknowledgement

The authors would like to express their thanks to Prof. Dr. *H. Roushdy* for this fruitfull discussions.

Received June 6, 1985

Accepted in revised form November 14, 1985

References

[1] *R. D. H. Chu*, Technical Report Series No. 205, IAEA, Vienna, 1981, 97

[2] *A. M. Eid, M. I. El Gohary, S. M. Kamal*, Proc. of the First National Conf. on Biophysical and Bioengineering Sciences, Cairo, 1982, 44

[3] *F. I. A. Said, G. Földiák, L. Wojnárovits*, Radiochem. Radioanal. Letters **32** (1978) 223

[4] *F. I. A. Said, G. Földiák, L. Wojnárovits*, Radiochem. Radioanal. Letters **34** (1978) 339

[5] *T. J. Sworsky*, Radiation Res. **4** (1956) 483

[6] *T. J. Hardwick*, Can. J. Chem. **30** (1952) 23

[7] *A. Miller, W. L. McLaughlin*, Int. J. Appl. Rad. Isotopes **33** (1982) 1299

[8] *M. J. Rossiter, J. W. Wood*, Technical Report Series No. 132, IAEA, Vienna, 1971

[9] *B. L. Gupta*, Technical Report Series No. 205, IAEA, Vienna, 1981,

[10] *A. Kovacs, V. Stenger*, IAEA-TECDOC-321, 1984, 117

[11] *K. H. Chadwick*, Technical Report Series No. 205, IAEA Vienna, 1981, 33

Isotopenpraxis 22 (1986) 11, S. 385 bis 388

Neutron Radiographic Detection of Surface Corrosion on Aluminium*)

J. Rant, R. Ilić

("J. Stefan Institute", "E. Kardelj", University of Ljubljana, Yugoslavia)[1])

G. Pregl

(University of Maribor, Maribor, Yugoslavia)

S. Demirović

("SOKO", Mostar, Yugoslavia)

The contrast sensitivity of neutron radiographic detection of corrosion products in aluminium is nearly one order of magnitude better than with soft (30 kV) X-ray radiography. With neutron radiography quantitative measurements of the depth of surface corrosion of aluminium objects can be effectively performed. Experimentally it has been confirmed that through 1 cm thick Al slab approximately 0.1 – 0.2 mm thick layer of corrosion products can easily be detected. Neutron radiography can thus be a valuable nondestructive method for the detection and quantitative evaluation of corrosion on aluminium structures. Aluminium honeycomb structures can thus be simultaneously inspected for corrosion nests, moisture and defects in adhesive layer.

Die Kontrastempfindlichkeit des neutronenradiographischen Nachweises von Korrosionsprodukten im Aluminium ist ungefähr eine Größenordnung besser als die mit weichen Röntgenstrahlen (30 kV). Mit Hilfe der Neutronenradiographie können quantitative Messungen der Tiefe einer Korrosion des Aluminiums effektiv durchgeführt werden. Es konnte experimentell bestätigt werden, daß die Korrosionsprodukte von ungefähr 0,1 ... 0,2 mm Dicke in einer 1 cm dicken Al-Platte leicht nachweisbar sind. Die Neutronenradiography kann somit als zerstörungsfreie Methode zum Nachweis und zur quantitativen Bestimmung einer Korrosion im Al benutzt werden. Wabenförmig strukturiertes Al kann gleichzeitig auf Korrosionsreste, Feuchtigkeit und auf Defekte von Haftschichten durchmustert werden.

Keywords

aluminium; corrosion; corrosion products; neutron radiography; nondestructive analysis; nondestructive testing

1. Introduction

The nondestructive detection of corrosion on aluminium structures is an important technical problem. To solve this problem various nondestructive inspection techniques have been developed and are in wide use today. However, detection of surface and intergranular corrosion on aluminium structures is for standard nondestructive inspection techniques such as X-ray radiography and ultrasonic methods a difficult problem, very often requiring partial or complete disassembly of the inspected objects, especially if this are screened or hidden behind thick structural elements. Due to its unique radiographic properties neutron radiography has been proposed and successfully used for detection and localisation of corrosion on various aluminium aircraft structural components [1]. In this communication we shall describe experiments we have performed in order to assess the possibilities of the method for quantitative determination of the depth of corrosion products on aluminium. In addition to the detection and localization of the corrosion quantitative neutron radiography would also help to determine the extent of the corrosion damage. Further, we were interested in determining the minimal thickness of realistic corrosion products which could be detected with our technique on aluminium objects.

2. Attenuation properties of aluminium and its corrosion products for thermal neutrons and soft X-rays

In many common cases, e.g. in the presence of halides in marine environment, the corrosion process on aluminium can be considered as the following transformation:

$$\text{Al(density } 2\cdot 7\ \text{g/cm}^3) \rightarrow \text{various Al trihydrates } (\text{Al(OH)}_3\text{-density } 2.4\ \text{g/cm}^3), \text{ monohydrates and Al salts.}$$

In practice $Al(OH)_3$ (hydrargilite or bayerite) constitutes 60 – 90% of corrosion products. In addition to trihydrates also monohydrates (boehmite, diaspore), oxides and various aluminium salts can be present. Moisture and water droplets can often accompany corrosion nests.

The attenuation properties of Al corrosion products can thus substantially differ from case to case depending on actual composition. Therefore for each typical case of inspection the expected actual composition of corrosion pro-

*) Presented at the XIVth International Symposium on "Autoradiography", November 19 – 23, 1984, Reinhardsbrunn, G.D.R.

[1]) Address: Yu-61111 Ljubljana, Jamova 39 P. O. Box 100

ducts has to be determined. Only as a rough approximation can one assume $Al(OH)_3$ to be the only constituent of Al corrosion products. Nevertheless, for illustrative purposes it is interesting to compare attenuation coefficients of base aluminium metal, of $Al(OH)_3$ and $AlO(OH)$ and water for thermal neutrons and soft X-rays (30–40 kV), which are often used for radiographic inspection of Al structures. The microscopic cross sections and attenuation coefficients of relevant elements and materials are presented in Tab. 1.

Tab. 1. Linear attenuation coefficients of thermal neutrons (Σ) and soft (30 kV) X-rays (μ) for aluminium, $Al(OH)_3$ and water

Element or material	microscopic cross section (barn) or linear attenuation coefficient (cm^{-1})	
	X-rays (30 kV)	Thermal neutrons
Al	49·3 barn/atom	1·42 barn/atom
O	9·81	4·20
H	0·597	38·34
Al	3·0 cm^{-1}	0·0861 cm^{-1}
$Al(OH)_3$	1·501	2·40
$AlO(OH)$	2·16	1·50
H_2O	0·368	2·706

On the basis of the above attenuation coefficients, we can conclude on the following radiographic properties of neutrons and soft X-rays for the detection at corrosion on aluminium:

- aluminium is almost transparent for neutrons and it is possible to inspect neutronographically even thick (several cm) Al objects. This is increasingly difficult and costly with soft X-rays. In this case, thick Al parts can easily mask structural features and loss of material due to corrosion.
- due to the high neutron scattering cross section of hydrogen the corrosion products are highly opaque to neutrons and thus can be detected even through thick Al slabs. In the case of soft X-rays the difference in the attenuation coefficients of aluminium and its corrosion products is smaller than with thermal neutrons and one canthus expect a smaller detection sensitivity.

In the assessment of the attenuation properties of aluminium corrosion products it is also important to consider the relation between the depth of corrosion (loss of Al base metal) and the thickness of corrosion products:

$$\left.\begin{array}{l} \text{1 g of corroded Al yields: } 2{\cdot}89 \text{ g of } Al(OH)_3 \text{ or} \\ 1\ cm^3 \text{ of corroded Al yields: } 3{\cdot}22\ cm^3 \text{ of } Al(OH)_3 \end{array}\right\} \quad (1)$$

Therefore in the case of $Al(OH)_3$ the thickness of the layer of corrosion products is ~ 3 times the depth of corroded base aluminium. Knowing the composition of corrosion products and measuring their thickness by suitable nondestructive method, e.q. from neutron radiographs, it is possible to estimate also the loss of the base aluminium metal.

3. Contrast sensitivity of radiographic detection of corrosion products for thermal neutrons and soft X-rays

A relative merit of neutron radiography against soft X-ray radiography for the detection of corrosion products on aluminium can be obtained by comparing the relative contrast K_{cp} of the image of an $Al(OH)_3$ inclusion imbedded in aluminium base metal:

$$K_{cp} = \frac{D_{cp} - D_{Al}}{D_{Al}} \quad (2)$$

where is:

D_{cp} – film density of the image of corrosion products
D_{Al} – film density of the image of uncorroded surrounding aluminium

Assuming a linear relationship between film density and film exposure, for the geometry of the simple radiographic inspection presented in Fig. 1 and for $x_{cp} \ll 1/\Sigma_{cp}$ (or $1/\mu_{cp}$) one obtains:

$$\left.\begin{array}{l} K_{cp}\ (\text{neutrons}) = 1 - e^{-\Sigma_{cp}x_{cp} + \Sigma_{Al}x_1} \simeq 3\Sigma_{cp}x_1 \\ K_{cp}\ (\text{X-rays}) = 1 - e^{-\mu_{cp}x_{cp} + \mu_{Al}x_1} \simeq (3\mu_{cp} - \mu_{Al})\, x_1 \end{array}\right\} \quad (3)$$

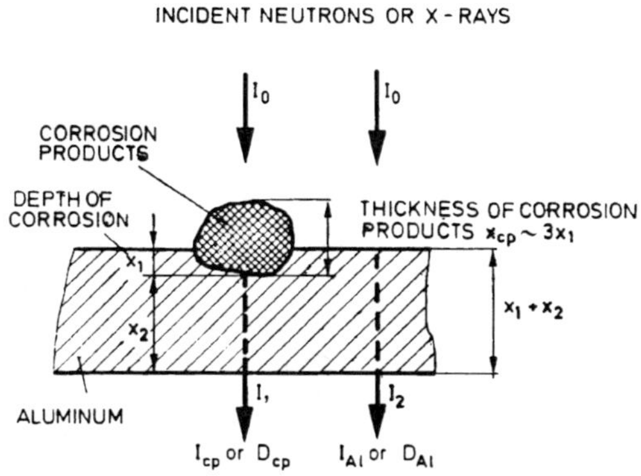

Fig. 1. Geometry of simple radiographic inspection of surface corrosion

Taking into account the approximate relation $x_{cp} \simeq 3x_1$ between the thickness of corrosion products and depth of corrosion (x_1), for the limiting case of inspection of very thin layers we obtain:

$$\frac{K_{cp}\,(\text{neutrons})}{K_{cp}\ (\text{X-rays})} \simeq \frac{3\Sigma_{cp}}{3\mu_{cp} - \mu_{Al}} = 4{\cdot}8 \quad (4)$$

The relative image contrast and therefore also the contrast sensitivity for detection of surface corrosion on aluminium is almost 5 times greater with thermal neutron radiography than with soft X-ray radiography.

Knowing the minimal visually detectable relative image contrast K_{cp}^{min} we can estimate the minimal detectable thickness X_{cp}^{min} of corrosion products:

$$\left.\begin{array}{l} x_{cp}^{min}\ (\text{neutrons}) = \dfrac{1}{\Sigma_{cp}} K_{cp}^{min} \\ x_{cp}^{min}\ (\text{X-rays}) = \dfrac{1}{\mu_{cp} - \frac{1}{3}\mu_{Al}} K_{cp}^{min} \end{array}\right\} \quad (5)$$

Assuming the minimal visually detectable relative image contrast K_{cp}^{min} conservatively to be 0·01, the minimal neutronographically detectable thickness of pure $Al(OH)_3$ is about 0·04 mm and for soft X-rays amounts to about 0·2 mm. In practice these figures can be considerably greater depending on the actual composition of the corrosion products.

4. Description of the experiment

The relation between the film density and the thickness of realistic corrosion products was experimentally determined from neutronoradiographs of aluminium test objects. In each 1 cm thick Al block two holes of equal diameter (sets of 1·0, 0·4, 0·2 and 0·1 cm diameter) and depth were drilled. For each set of holes the hole depth ranged from 0·01 to 0·8 cm. One of the holes was filled with corrosion products, the other left empty and the Al block covered with thin (0·02 mm) Al foil. The corrosion products were obtained by scraping Al pieces corroded in sea water, drying in open air, grinding and sieving. The mass thickness M_{cp}/S of the corrosion products in each hole was determined by precision weighting and the tap density (g/cm³) of the filled powder ranged from 1·1 to 2·0 g per cm³, significantly lower than the value 2·4 g/cm³ for pure $Al(OH)_3$. Each set of Al blocks was simultaneously neutron radiographed in the thermal column NR facility of the Ljubljana TRIGA Mark II reactor, described previously [3]. A direct neutron radiographic method with Gd (0·025 mm) screen and Kodak M S.C. film was used.

The neutron radiographs (positive prints) of the 1 cm hole set are presented in Fig. 2. The hole depths and mass thickness of corrosion products are quoted below each hole. Background film density due to gamma-rays was negligible. Note the reversed contrast between the hole filled with corrosion products (dark on the positive print due to higher neutron attenuation) and the image of the empty hole (light) in comparison with the image density of the aluminium base metal.

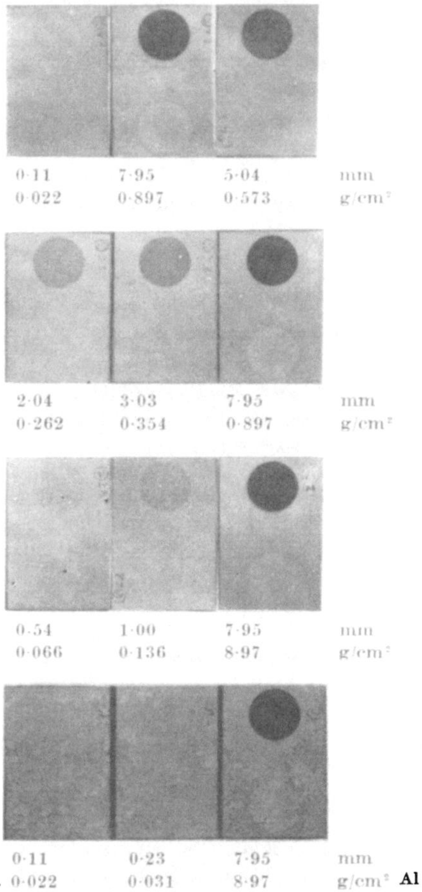

Fig. 2. Al test block (1 cm thick) with holes (∅ 10 mm) filled with corrosion products and unfilled holes. Note the reverse contrast between filled (darker) and unfilled (lighter) holes in comparison with the image of Al base metal. All filled holes are visible.

The calibration curves relating net density of neutronoradiographs and the thickness of corrosion products was been measured. The densitometric readings were taken in the middle of the filled hole (D_{cp}), in the middle of the empty hole (D_H) and in the middle of the Al block (D_{Al}). The precision of the densitometric readings was only 5–10%. It is easy to show that the graphs of log (D_{cp}/D_H) and log (D_{cp}/D_{Al}) are linear functions of the mass thickness of the corrosion products. Experimentally obtained graphs of log (D_{cp}/D_H) and log (D_{cp}/D_{Al}) for 1·0 cm holes are presented in Fig. 3. From the slopes of these linear functions estimates of Σ_{cp} and $\Sigma_{cp} - \Sigma_{Al}$ can be obtained respectively. In our case the estimated value for Σ_{cp} (0·55 cm⁻¹) is significantly lower than the value (1·2 cm⁻¹) for pure $Al(OH)_3$ at the tap density 1·25 g/cm³. This indicates the presence of monohydrates and other aluminium salts in addition to $Al(OH)_3$. In our particular case with accurate densitometry of neutron radiographs the minimal thickness, which could be reliably determined is at least 0·1 mm (mass thickness 0·02 g/cm²). Knowing the composition, it could be possible to determine the thinning of the Al base metal.

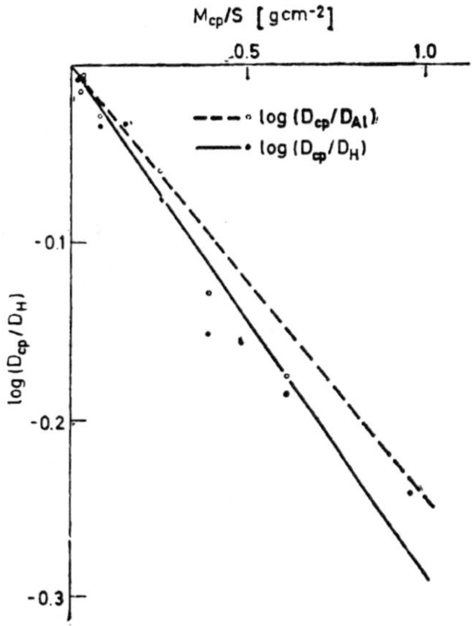

Fig. 3. Calibration curves relating net film density D_{cp} to the thickness of the layer of corrosion products

5. Applications

The neutron radiographic detection of surface corrosion on aluminium structures has been found useful in the maintenance of aircraft, especially those exposed to marine environments. A mobile ^{252}Cf neutron radiographic facility has been reported in use for this purpose [1]. Neutron radiography enabled the inspection of inaccessible or masked aluminium structures, requiring long and costly dissassembly of parts. A reduction of inspection time to about 15% in comparison with visual and common NDT techniques is reported. This can significantly increase the availability of costly modern aircrafts and improve their safe exploitation. As an example, in Fig. 4 we present a neutron radiograph of an Al clad honeycomb structure, artificially aged to contain moisture and corrosion nests. Simultaneously it is possible to observe surface corrosion, air bubbles

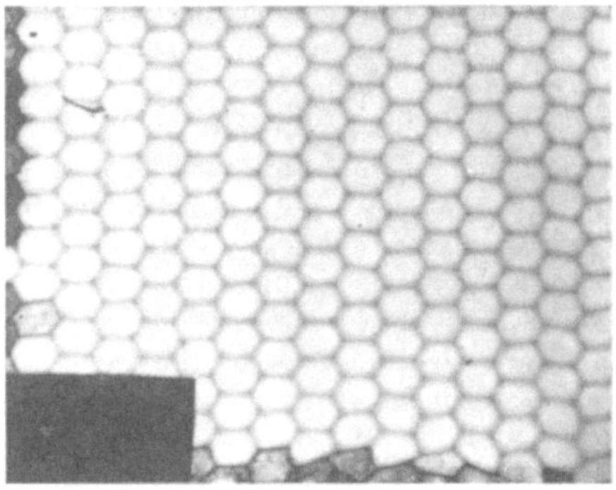

Fig. 4. Neutron radiograph of artificially aged aluminium clad honeycomb structure

in the glue and distribution of glue. With neutron radiography the sensitivity of detecting surface corrosion, moisture or defects in the glue distribution should not be inpaired by varying the thickness of Al structure.

6. Conclusions

The results of our study can be summarized as follows:

- with neutron radiography it is possible to visualize directly surface corrosion on Al objects, while with soft X-rays one detects variations in the thickness of the Al base,
- the inspection capability of neutron radiography for detecting surface corrosion is practically almost insensitive to the variations in the thickness of the Al base metal. Several cm thick Al structures can easily be inspected,
- the minimal thickness of corrosion products one can neutronographically detect on aluminium depends on the amount of $Al(OH)_3$ and $AlO(OH)$ in their composition, and in our case was found to be about 0·022 g/cm³,
- using previously calibrated measurements it is possible from neutron radiographs to estimate the thickness of corrosion products.

Received October 15, 1985

References

[1] *J. John*, Neutron radiography for Nondestructive Testing (Proc. Conf. "Nondestructive Evaluation of Materials", ed. by *J. J. Burke*, *V. Weis*) New York: Plenum Press 1979, 151 – 182

[2] *M. R. Hawkesworth*, Atomic Energy Review **15** (1977) 169

[3] *J. Rant* et al., Proc. Nat. Conf. on Utilization of Nuclear Research Reactors in Yugoslavia. Belgrade 1978, 505 – 513

Isotopenpraxis 22 (1986) 11, S. 388 bis 392

Bestimmung der Isotopenverteilung in ^{14}C-markierten Kohlenwasserstoffen durch thermische Fragmentierung

F.-D. Kopinke

(Akademie der Wissenschaften der DDR, Zentralinstitut für Organische Chemie, Bereich Organische Grundstoffe, Leipzig)[1])

J. Dermietzel, W. Jockisch, G. Räuber

(Akademie der Wissenschaften der DDR, Zentralinstitut für Isotopen- und Strahlenforschung)[1])

Die radiogaschromatographische Analyse der Pyrolyseprodukte von geeigneten markierten Kohlenwasserstoffen gestattet eine eindeutige und quantitative Bestimmung der ^{14}C-Verteilung in diesen Verbindungen. In der vorliegenden Arbeit wird eine einfache, schnelle und vielseitig anwendbare experimentelle Methode dafür beschrieben.

The radio gas chromatographic analysis of the pyrolysis products of suitable labelled hydrocarbons allows an unambiguous and quantitative determination of the ^{14}C-distribution in these compounds. In this paper an uncomplicated, prompt and versatilly applicable experimental method is described for it.

Keywords

carbon 14; gas chromatography; hydrocarbons; labelled compounds; pyrolysis; tracer techniques

1. Einführung

Für eine C-Markierung organischer Verbindungen kommen prinzipiell drei Isotope in Betracht: ^{11}C, ^{13}C und ^{14}C. Benötigt man Angaben über die Markierungsposition innerhalb eines Produktmoleküls, was für kinetisch-mechanistische Untersuchungen häufig der Fall ist, so bietet das stabile ^{13}C-Isotop Vorteile, weil seine Verteilung mit Hilfe der ^{13}C-NMR-Spektroskopie direkt zugänglich ist, während für eine Positionsanalyse ^{14}C-markierter Verbindungen keine vergleichbaren spektroskopischen Methoden existieren. Die Anwendung des ^{13}C-Isotops als Tracer ist jedoch aus Gründen der Verfügbarkeit (Preis!), der Nachweisempfindlichkeit und des meßtechnischen Aufwandes nur begrenzt möglich.

Für eine Positionsanalyse ^{14}C-markierter Verbindungen sind meist recht aufwendige chemische Abbauverfahren

[1]) Anschrift: DDR-7050 Leipzig, Permoserstraße 15

erforderlich [1], deren Übertragung auf den Halbmikro- oder Mikromaßstab zusätzliche Schwierigkeiten bereiten kann. Aufbauend auf langjährige Erfahrungen auf dem Gebiet der thermischen Kohlenwasserstoffwandlung im Bereich Organische Grundstoffe des Zentralinstitutes für Organische Chemie und auf dem Gebiet der radiogaschromatographischen Analyse im Bereich Tracerforschung des Zentralinstitutes für Isotopen- und Strahlenforschung wurde die bekannte Methode der Pyrolyse-GC [2, 3] dahingehend weiterentwickelt, daß durch Kombination von präparativer GC, thermischer Fragmentierung und Radio-GC eine schnelle Positionsanalyse für zahlreiche Verbindungen möglich wird, selbst wenn diese nicht in reiner Form, sondern als Bestandteile von Produktgemischen vorliegen.

2. Experimentelle Grundlagen

Das Schema der Versuchsapparatur ist in Abb. 1 dargestellt. Sie besteht aus einem Gaschromatographen GCHF 18.3 (Fa. Chromatron Berlin) mit WLD, einem Durchfluß-Proportionalzählrohr (Eigenbau, 10 ml Volumen, 30 µm Anodendraht, Propan:$H_2 \approx 1:1$, Arbeitspannung 2,5 kV) mit nachgeschalteter Elektronik und Registriereinheiten (Drucker und Impulsdichteschreiber), zwei Probengabeventilen, einer Kühlfalle und einem elektrisch beheizten Pyrolysereaktor (3 × 0,5 × 200 mm gewendeltes Rohr aus Edelstahl X10 CrNiTi 18:9). Beide Probengabeventile, das Durchflußzählrohr und die sie verbindenden Stahlkapillaren sind auf Temperaturen bis 150 °C beheizbar.

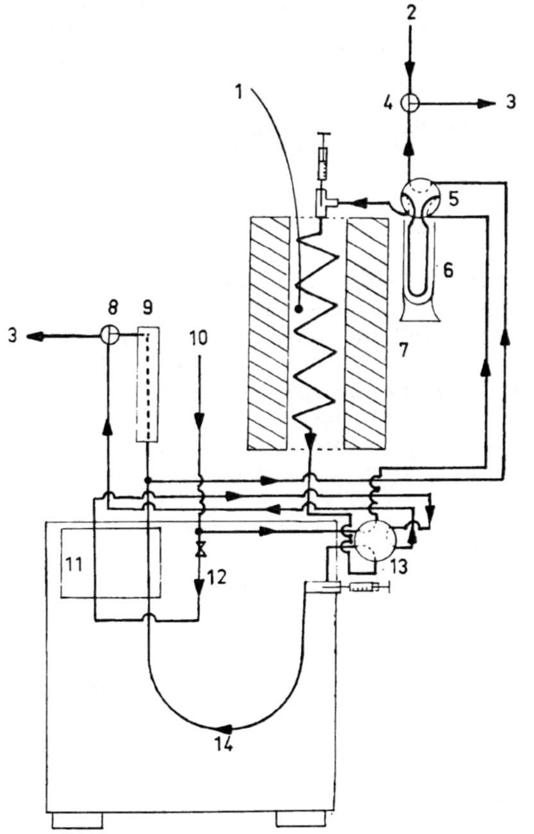

Abb. 1. Apparatur zur Pyrolyse und RGC von ^{14}C-markierten Verbindungen

1 – Thermoelement; 2 – Zählgas; 3 – Abgas; 4, 8 – Dreiwegehähne; 5, 13 – Probengabeventile; 6 – Kühlfalle; 7 – Pyrolysereaktor; 9 – Durchflußproportionalzählrohr; 10 – H_2-Trägergas; 11 – WLD; 12 – H_2-Vergleichsgas; 14 – Trennsäule

Die Apparatur erfüllt folgende Funktionen:

- GC-Trennung eines Substanzgemisches,
- Isolierung der interessierenden Komponente,
- thermische Zersetzung dieser Komponente und
- Radio-GC der Pyrolyseprodukte.

Der Versuchsablauf beginnt mit der Dosierung einer die interessierende Verbindung enthaltenden Probe (0,5 bis 10 µl) über den Verdampfer des GC auf eine analytische oder präparative Trennsäule. Bei diesem ersten Schritt wird an Stelle des Zählgases (Propan) der H_2-Vergleichsgasstrom in umgekehrter Richtung durch das Zählrohr geleitet. Die den WLD nach erfolgter chromatographischer Trennung verlassenden Komponenten gelangen so zu einem Probengabeventil (5), welches das Ausfrieren der interessierenden Komponente in reiner Form in einer Kühlfalle (−196 °C, leeres Edelstahlrohr) ermöglicht.

Nach beendeter Vortrennung wird das Zählgas-Trägergasgemisch durch das Zählrohr geleitet (Umschalten der Dreiwegehähne 4 und 8), der Kühlfalleninhalt durch schnelles Erhitzen mit einem Wasserbad über den Pyrolysereaktor (Verweilzeit ≈ 1 s) auf die Trennsäule dosiert und das Gemisch der Pyrolyseprodukte radiogaschromatographisch analysiert. Liegt die interessierende markierte Verbindung schon in reiner Form vor, kann die Vortrennung entfallen und die Probendosierung direkt am Reaktoreingang erfolgen. Das zweite Probengabeventil (13) dient der Trennung von Pyrolysereaktor und Radio-GC, wodurch ein unabhängiger Betrieb beider Teile ermöglicht wird.

3. Mechanistische Aspekte des thermischen Zerfalls von Kohlenwasserstoffen

Als Mechanismus für den thermischen Zerfall von Kohlenwasserstoffen im hier interessierenden Temperaturbereich > 600 °C wird heute eine unverzweigte Radikalkettenreaktion mit kinetischen Kettenlängen von ca. 10^1 bis 10^2 allgemein akzeptiert [4]. Das äußerst komplexe Reaktionsgeschehen führt zur Bildung von Spaltprodukten und sogenannten Aufbauprodukten. Erstere sind im Sinne dieser Arbeit erwünscht, letztere sind unerwünscht und verursachen u. a. wegen ihrer Neigung zur Bildung fester koksähnlicher Ablagerungen in Transferleitungen Schwierigkeiten.

Voraussetzung für die Ausnutzung von Pyrolysereaktionen für eine Positionsanalyse ist der eindeutige Zusammenhang zwischen der Beteiligung von C-Atomen an bestimmten Pyrolyseprodukten und ihrer Position im Ausgangsmolekül. Ein solcher eindeutiger Zusammenhang ist für aliphatische Kohlenwasserstoffe > C_4 i. a. nicht gegeben, selbst wenn es gelingt, durch niedrige Umsetzungsgrade eine Aktivitätsumverteilung als Folge von Sekundärreaktionen weitgehend zu vermeiden. In Abb. 2 sind einige Beispiele aus eigenen Arbeiten für z. T. unerwartete Aktivitätsverteilungen in den Spaltprodukten von markierten Modellkohlenwasserstoffen dargestellt, die nur durch Radikalisomerisierungen oder sigmatrope Umlagerungen erklärbar sind. Aus der Aktivitätsverteilung in den Spaltprodukten kann daher nicht in einfacher Weise quantitativ auf die ursprüngliche Aktivitätsverteilung in der markierten Ausgangsverbindung geschlossen werden. Qualitative Aussagen oder halbquantitative Abschätzungen sind in

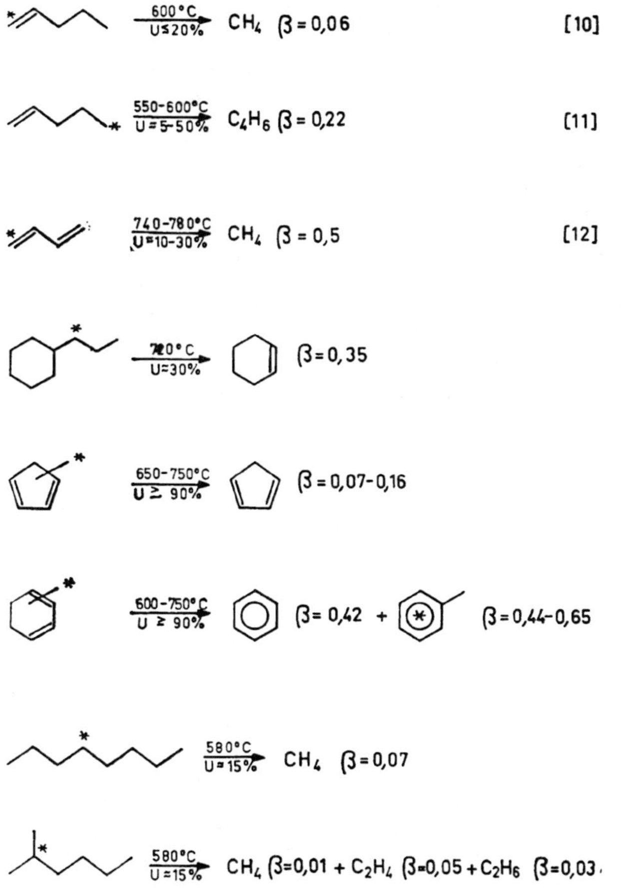

Abb. 2. Beispiele für spezifische und unspezifische Aktivitätsverteilung in den Pyrolyseprodukten ^{14}C-markierter Kohlenwasserstoffe (β ... relative molare Aktivität)

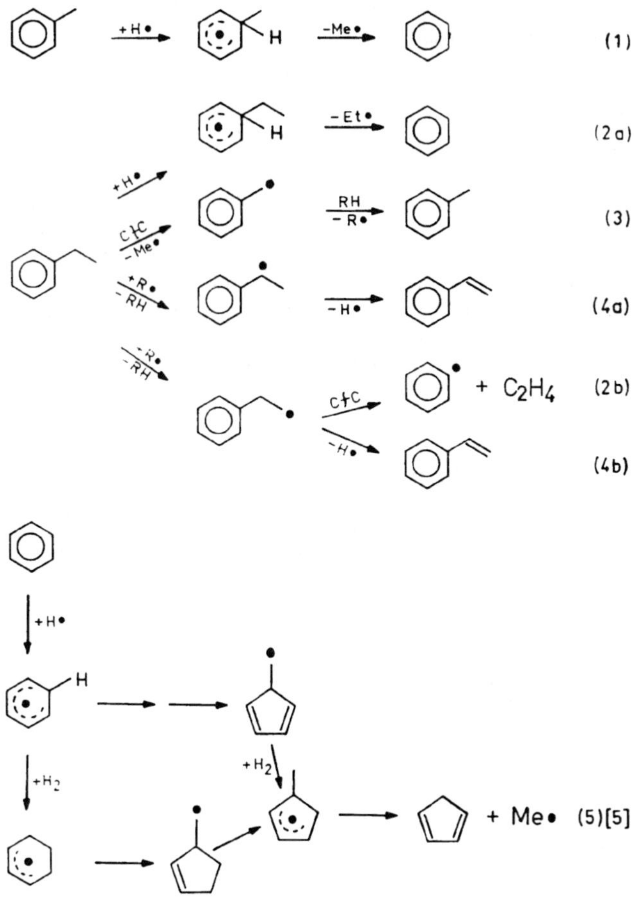

Abb. 3. Thermische Fragmentierung von Alkylbenzenen in Wasserstoffatmosphäre nach dem Radikal-Ketten-Mechanismus

solchen Fällen dennoch möglich, wenn man über eindeutig spezifisch markierte Testsubstanzen verfügt.

Bessere Voraussetzungen für einen definierten Abbau als aliphatische Kohlenwasserstoffe besitzen Alkylbenzene. Der thermisch sehr stabile aromatische Ring bleibt weitgehend erhalten und übt einen starken dirigierenden Einfluß auf die Reaktionen am Alkylsubstituenten aus (Abb. 3). Hauptprodukte der Radikalreaktionen sind stets Benzen, Toluen und Styren. Die Entalkylierung des Benzenringes wird durch einen hohen Anteil von H-Atomen als kettentragende Radikale begünstigt, was man durch Verwendung von Wasserstoff als Trägergas erreichen kann ($R\cdot + H_2 \rightarrow RH + H\cdot$). Allerdings muß man unter diesen Bedingungen einen geringen, aber signifikanten Anteil an Ringspaltungsreaktionen in Kauf nehmen, die heterogen über die Hydrierung von Koks ($Ph - R \rightarrow C \xrightarrow[+H_2]{\text{langsam}} CH_4$) oder homogen über den Zerfall des Cyclohexadienylradikals (Gl. 5 in Abb. 3) verlaufen können.

Die Pyrolyse von Alkylbenzenen erlaubt prinzipiell keine Positionsanalyse innerhalb des aromatischen Ringes; dieser wird deshalb als Ganzes betrachtet.

Von entscheidender Bedeutung für die Brauchbarkeit der thermischen Fragmentierung von Alkylbenzenen ist die Frage nach dem Ausmaß eines möglichen C-Scramblings, das für diese Moleküle unseres Wissens bisher nur unter extremen Bedingungen [6], als Radikalkationen in der MS [7] oder ausgehend von exponierten Strukturen, die zur Ausbildung von Carbenen führen können (z.B. Benzocyclobuten [8]), beobachtet wurde.

4. Analytik

Die Fragmentierung von Alkylbenzenen liefert zwei unabhängige Informationen über die Markierungsposition im Molekül — als Aktivitätsverteilung in den aromatischen und in den nichtaromatischen Fragmenten. Führt man alle gaschromatographischen Trennungen auf nur einer gepackten Trennsäule aus (Abb. 1), wird man sich meist mit einer summarischen Erfassung der C_3^--Fraktion bescheiden müssen. Die Auswahl der Trennsäule wird durch die Zusammensetzung des Gemisches, aus dem die interessierenden Alkylbenzene abgetrennt werden sollen, bestimmt. Eine Auftrennung der Pyrolyseprodukte in C_3^-, Benzen, Toluen, Xylene oder Ethylbenzen/Styren und längerkettige Alkylbenzene ist an vielen unpolaren oder polaren Trennphasen leicht möglich.

In der vorliegenden Arbeit kam eine isotherm (40 bis 90 °C) betriebene 2 m × 4 mm SE52-Säule (5% auf Chromatron N Super) zur Anwendung.

5. Ergebnisse

In Tabelle 1 sind relative molare Aktivitäten der wesentlichen aromatischen Pyrolyseprodukte aus einigen reinen, spezifisch ^{14}C-markierten Alkylbenzenen zusammengestellt. Die Fehlerangaben sind mittlere Abweichungen des Einzelwertes vom Mittelwert aus je drei Pyrolyseversuchen. Die erhaltenen Werte entsprechen innerhalb eines für manu-

Tab. 1. Relative molare Aktivitäten von Pyrolyseprodukten aus ^{14}C-markierten Alkylbenzenen (α = 2 – 5 kBq mg^{-1})

Einsatzprodukt	$T_{Pyrolyse}$	U_{EP} [1])	Pyrolyseprodukte			C_3^- [2])
	[°C]	[%]	Benzen	Toluen	Styren	[AKt-%]
[Ring-^{14}C]-Toluen	810	30	1,02 ± 0,05	1,00	–	$\lesssim$ 2,0
[7-^{14}C]-Toluen			< 0,003	1,00		
[Ring-^{14}C]-Xylene	750	50	1,00 ± 0,10	0,97 ± 0,06		$\lesssim$ 1,5
[7-^{14}C]-Xylene			< 0,01	0,49 ± 0,02		–
[Ring-^{14}C]-Ethylbenzen	700	70	1,05 ± 0,05	0,97 ± 0,07	0,98 ± 0,06	$\lesssim$ 0,5
[7-^{14}C]-Ethylbenzen			< 0,005	0,98 ± 0,09	1,06 ± 0,05	–
[8-^{14}C]-Ethylbenzen			< 0,005	< 0,01	1,00 ± 0,05	–
[7-^{14}C]-Propylbenzen	650	70	< 0,01	0,94 ± 0,07	1,02 ± 0,02[3])	–

[1]) Umsetzungsgrade auf der Basis der detektierten Spaltprodukte; tatsächliche Umsetzungsgrade sind jedoch höher, da Bildung hochsiedender Aufbauprodukte

[2]) Angaben in AKt-% von der dosierten Probe

[3]) Styren + Ethylbenzen

elle Peakflächenauswertung (Dreiecksmethode) typischen Fehlerintervalls ($\leqq$ 10%) dem mechanistisch begründeten Fragmentierungsschema (Abb. 3). Ein signifikantes C-Scrambling (> 1%) zwischen sp^2- und sp^3-hybridisierten C-Atomen kann ausgeschlossen werden. Ebenso kann man aus dem β_{rel}-Wert des Toluens, das durch Entmethylierung von (Me-^{14}C)-Xylenen gebildet wird (0,49 ± 0,02), auf einen vernachlässigbar geringen kinetischen Isotopeneffekt bei der Spaltung einer ^{12}C-^{14}C-Bindung unter diesen Bedingungen schließen.

Der Anteil von Ring-C-Atomen, die bei der Pyrolyse in nichtaromatische Spaltprodukte (bevorzugt Methan) umgewandelt werden, steigt mit der Reaktionstemperatur an.

Als Mittel zur Minimierung des heterogenen Beitrages zu dieser unerwünschten Nebenreaktion hat sich die Schwefelung der metallischen Reaktoroberfläche als sehr wirksam erwiesen. Zu diesem Zweck haben wir ca. 10 µl CS_2 so in den vom Vergleichsstrom durchspülten Reaktor (T_R = 650 bis 800 °C) dosiert, daß das gebildete H_2S nicht über die Trennsäule, sondern direkt in die Abgasleitung gelangt. Diese nur wenige Minuten beanspruchende Oberflächenpassivierung muß in bestimmten Abständen wiederholt werden, weil die Metalloberfläche im H_2-Strom reaktiviert wird (erkennbar am Tailing des Methanpeaks durch langsame Hydrierung von Koks). Der Einsatz eines Quarzreaktors wäre unter diesem Gesichtspunkt sicher wünschenswert, doch stehen dem konstruktive Erschwernisse gegenüber.

Die Auswirkungen der homogen und heterogen verlaufenden Ringspaltung bestehen darin, daß kleine Aktivitätsanteile in der Seitenkette nur unsicher nachgewiesen werden können, da die C_3^--Fraktion mit einem „Aktivitätsblindwert" behaftet ist ($\lessapprox$ 0,5 – 2 AKt-%) und die relativen molaren Aktivitäten der aromatischen Fragmente in diesem Fall nur wenig von 1,0 abweichen würden (Differenz zweier großer Zahlen ist ungenau).

Kleine Aktivitätsanteile im aromatischen Ring sind hingegen empfindlich nachweisbar. Die Nachweisgrenze wird hier lediglich durch die spezifische Aktivität der Probe und die Empfindlichkeit des Aktivitätsdetektors begrenzt. Versuche zur Senkung der Reaktionstemperatur durch Zusatz von Initiatoren, die die Radikal-Ketten-Reaktion starten (z. B. Aceton), führten nur zu geringen Effekten.

Neben der thermischen Entmethylierung im Temperaturbereich 750 – 850 °C ist in Analogie zu großtechnischen Verfahrensvarianten eine katalytische Entmethylierung von Methylaromaten möglich, doch entstehen auch unter diesen Bedingungen (T < 650 °C) meist geringe Mengen nichtaromatischer Spaltprodukte.

Eine Ausweitung des Anwendungsbereiches der vorgestellten Methode auf Alkylcyclohexane erscheint möglich durch Einbau einer der thermischen Fragmentierung vorgelagerten Dehydrierstufe (Pt, 400 °C).

Der klassische Weg zur Positionsanalyse von ^{14}C-markierten Methylaromaten ist ihre Oxidation zu Arylcarbonsäuren und deren katalytische Decarboxylierung, z. B.

$$\mathrm{Ar-CH_3 \xrightarrow[Na_2CO_3/H_2O]{KMnO_4} Ar-COOH \xrightarrow[CuO]{Chinolin} ArH + CO_2}$$

Santrock und *Hayes* [9] haben kürzlich die pyrolytische Decarboxylierung (500 – 600 °C, 5 – 20 min) von aromatischen Carbonsäuren in Verbindung mit einer MS-Analyse der Isotopenzusammensetzung (C und O) beschrieben.

Wir haben die thermische Decarboxylierung von ^{14}C-Benzoesäure (als 25%ige methanolische Lösung) in der Impulspyrolyseapparatur nach Abb. 1 untersucht. Bei Reaktionstemperaturen von 700 – 750 °C decarboxyliert Benzoesäure glatt und quantitativ (Verweilzeit $\approx$ 1 s). Ein signifikantes C-Scrambling und die Hydrogenolyse des aromatischen Ringes (> 3%) konnten durch Prüfung mit reiner (Ring-^{14}C)-Benzoesäure ausgeschlossen werden.

Für die Ermittlung der ^{14}C-Verteilung zwischen Ring- und Carboxylgruppe ist keinerlei Peakflächenbestimmung, sondern lediglich das Verhältnis der im CO_2- und im Benzenpeak registrierten Impulse erforderlich (A_{CO_2}/A_{Benzen}). Eine Überprüfung der Richtigkeit der so erhaltenen ^{14}C-Verteilung in definierten Benzoesäuregemischen lieferte befriedigende Übereinstimmung (±5% bei α_{BS} = 0,1 bis 1 kBq mg^{-1}) mit den Ergebnissen der Chinolin/CuO-Methode [1]. Der Zeitaufwand für die Durchführung und Auswertung einer Decarboxylierung mit Hilfe der Pyrolyse-RGC beträgt ca. 10 min gegenüber mehreren Stunden nach dem klassischen Verfahren, was den Vorteil der von uns beschriebenen Methode unterstreicht.

Eingegangen am 16. Dezember 1985

Literatur

[1] *H. Simon, H. G. Floss*, Bestimmung der Isotopenverteilung in markierten Verbindungen, Springer-Verlag, Berlin 1967

[2] *E. Leibnitz, H. G. Struppe*, Handbuch der Gaschromatographie, 3. Aufl., Akadem. Verlagsgesellschaft Geest u. Portig K. G. Leipzig 1984, S. 599 ff.

[3] *V. Kozlik, V. Pacakova*, Chromatographia **14** (1981) 417 und **11** (1978) 5

[4] *G. Zimmermann, B. Ondruschka*, Mitteilungsbl. Chem. Gesellschaft **30** (1983) 73

[5] *R. Louw, J. H. M. Dijks, P. Mulder*, Recl. Trav. Chim. Pays-Bas **103** (1984) 271

[6] *R. D. Smith*, J. Phys. Chem. **83** (1979) 1553

[7] *J. Grotemeyer, H.-F. Grützmacher*, Organic Mass Spectr. **17** (1982) 353

[8] *W. S. Trahanovsky, M. E. Scribner*, J. Amer. Chem. Soc. **106** (1984) 7976

[9] *J. Santrock, J. M. Hayes*, Anal. Chem. 57 (1985) 1441

[10] *F.-D. Kopinke, G. Bach, B. Ondruschka, G. Zimmermann*, J. prakt. Chem. **325** (1983) 283

[11] *F.-D. Kopinke, G. Zimmermann, G. Bach, B. Ondruschka*, Int. J. Chem. Kinet. im Druck

[12] *F.-D. Kopinke, B. Ondruschka, G. Zimmermann*, Tetrahedron Lett. **24** (1983) 869

Isotopenpraxis 22 (1986) 11, S. 392 bis 395

Die Bestimmung der Tiefe der flüssigen Phase beim Strangguß*)

J. Hons

(Výzkumný ústav hutnictví železa — Forschungsinstitut für Eisenhüttenwesen)[1]

Die Bestimmung der Sumpftiefe wird auf einer Radialstranggußanlage für Stahl mit einem Format von 200 × 200 mm durchgeführt. Es werden eine größere Zahl von Punktstrahlern, die in hochschmelzenden Kapseln untergebracht sind, und ein weiterer Strahler, der sich in einer Bleikapsel befindet, zugesetzt. Die Lagekontrolle der Punktstrahler bei der Bewegung des Strangs erfolgt mit zwei Meßgeräten. Die genaue Tiefe der flüssigen Phase wird an Hand des tiefsten Strahlers im Strang unter Hinzurechnung derjenigen Stranglänge, die durch das markierte Blei gekennzeichnet ist, bestimmt. Die grundlegenden Untersuchungen wurden mit dem Gammaspektrometer TRACOR durchgeführt.

The crater depth is determined on a radial continuous casting machine for steel of the size 200 mm × 200 mm. Several point sources placed in refractory capsules and an additional source enclosed in a capsule of lead are added. The position of the point sources during the movement of the ingot is determined by the use of tow measuring devices. The exact depth of liquid phase is defined by the source located at the deepest level plus that ingot length indicated by the labelled lead. The studies were carried out by the use of the gamma spectrometer type TRACOR.

Keywords

autoradiography; casting; gamma spectrometers; radiation sources; steels; tracer techniques

Die Tiefe der flüssigen Phase des Stahls im Gießstrang unterhalb des Kristallisators gehört zu den wichtigsten Faktoren, die die Qualität des Erzeugnisses und die Sicherheit der Gießanlage bedingen. Das Messen dieser Größe ist sehr schwierig. Eine der älteren Methoden ist die Bestimmung von zugesetztem Blei in der Mikrostruktur des Stahls durch chemische oder metallographische Analyse. Zu den modernen Methoden gehört die Verwendung von Radionukliden (z.B. ^{35}S oder ^{32}P) und deren Nachweis mittels Autoradiographie. In diese Gruppe ist auch die Applikation von Gamma-Punktstrahlern und deren radiometrische Lokalisierung im Gießstrang mittels Geiger-Müller- oder Szintillationsdetektoren einzuordnen. Jede der genannten Methoden hat ihre Vor- und Nachteile hinsichtlich der Empfindlichkeit, Genauigkeit, der hygienischen und Arbeitsbedingungen und nicht zuletzt der unmittelbar praktischen Anwendbarkeit im Gießbetrieb. Der überwiegende Teil der bisherigen Arbeiten bezog sich auf den vertikalen Guß, bei dem sich der ferrostatische Druck auf die Stahlkristallisation und die Indikatorbewegung auswirkt. Schwieriger ist die technologische Lösung bei Radial- oder Horizontal-Gießanlagen, bei denen die mechanische und thermische Beanspruchung der Gußhaut am Umfang des Gießstranges nicht gleichmäßig ist, wodurch die Durchbruchgefahr erhöht wird. Die Wirkung des ferrostatischen Drucks ist hier geringer.

Die Aufgabe unseres Radioisotopenlabors bestand darin, ein für eine schnelle Betriebskontrolle geeignetes Verfahren zur genauen Bestimmung der Tiefe der flüssigen Phase an eine Radialstranggußanlage zu entwickeln und im Betrieb zu erproben.

Die experimentellen Arbeiten wurden an der Betriebsanlage des Šverma-Eisenwerkes in der Slowakei durchgeführt. Diese Radialstranggußanlage arbeitet mit vier Strängen; der Radius der Rundung beträgt 9 m, das Strangformat 200 × 200 mm. Bei der Lösung gingen wir davon aus, den Dichteunterschied zwischen Festkörper und Schmelze bei der Bewegung eines Festkörpers in der Schmelze zu nutzen. Gleichzeitig mußte die Gefahr des Festhaltens des Indikators in den Dendriten der Gußhaut besonders in der Rundung des Gießstranges und die des sofortigen Aufschmelzens des Indikators beachtet werden. Zur Verminderung dieser Einflüsse haben wir uns für eine kombinierte Methode der Markierung mit einer größeren Zahl von Strahlern entschieden. Für eine orientierende Kontrolle der Lage der Punktstrahler während des Gusses wurde eine kontinuierliche Registrierung der lokalen Radioaktivität verwendet. Die Hauptkontrolle wurde mittels autoradiographischer und gammaspektrometrischer Analyse an ausgewählten Bereichen des abgekühlten Stranges vorgenommen.

Die Tab. 1 gibt einen Überblick über die Haupteigenschaften der benutzten Radionuklide. Fünf von ihnen wurden in beiden Betriebsexperimenten als Punktstrahler eingesetzt. Das Radionuklid ^{124}Sb wurde einer der Kapseln zur Volumenmarkierung des Bleis beigegeben. Diese Kapsel bestand aus Blei. Die Metallkapseln, in denen die Punktstrahler eingeschlossen waren, sollten schwer sein und damit schnell im flüssigen Schmelzsumpf absinken, und sie durften nicht aufschmelzen, um den Übergang

*) Vortrag, gehalten auf dem 7. Kolloquium „Technische Anwendung radioaktiver Isotope im Berg- und Hüttenwesen", 13./14. 3. 1985, Hetzdorf b. Freiberg, DDR

[1]) Anschrift: Dobrá (ČSSR)

Tab. 1. Eigenschaften der verwendeten Radionuklide und Kapselwerkstoffe

	Element	Schmelzpunkt in °C	Halbwertszeit	Dichte in g/cm³	Form, Abmessung in mm
Radionuklid	^{46}Sc	1540	83,9 d	2,99	AlSc Gran.
	^{60}Co	1495	5,3 a	8,90	Granalien
	^{110m}Ag	960,5	250,4 d	10,49	Folie
	^{124}Sb	630,3	60,2 d	6,62	Granalien
	^{181}Hf	2222	49,0 d	13,29	Pulver
	^{182}Ta	2996	115,0 d	16,60	Pulver
Kapselwerkstoff	Mo	2622	–	10,22	10 × 20 ∅
	Mo*)	2622	–	10,22	10 × 10 ∅
	Stahl	1536 (Fe)	–	7,87	25 × 40 ∅
	Pb	327,4	–	11,34	32 × 70 ∅

(nach *Hodgman*, Chem. Tabellen)

*) andere Kapselgröße

der Punktstrahler in die flüssige Phase zu verhindern. Gleichzeitig wurde das Verhalten der Kapseln aus Metallen verschiedener Dichte und Schmelzpunkte und der unterschiedlichen Radionuklide geprüft. Die Aktivität der einzelnen Strahler lag zwischen 0,1 und 0,15 GBq. Die Kapseln hatten eine innere Öffnung von 4 mm Durchmesser mit Schraubverschluß. In der Mitte der mit Sb markierten Bleikapsel wurde eine kleine Molybdänkapsel mit verschiedenen Punktstrahlern untergebracht. In einem Falle handelte es sich um das Radionuklid ^{181}Hf, im anderen um ^{60}Co.

Abb. 1 zeigt das Schema der Gießanlage. Zwei Gerätesätze mit NaI(Tl)-Szintillationsdetektoren zum Messen und zur stetigen Registrierung der Gesamtimpulsrate wurden im Übergang der Rundung des Gießstrangs in die Horizontale installiert. Die Entfernung der ersten Sonde vom Stahlspiegel im Kristallisator betrug 10,8 m, die der zweiten Sonde 11,55 m. Die Detektoren wurden in wassergekühlten Kästen untergebracht. Der Abstand der Detektoren zur Strangoberfläche betrug 10 cm. Die Geräte und Kabel mußten mit einer Aluminium-Asbest-Abschirmung gegen Strahlungswärme geschützt werden.

Entsprechend dem Zeitprogramm wurden die in den Kapseln befindlichen Punktstrahler in den Kristallisator auf den schlackenlosen Stahlspiegel geworfen. Die Zeit ihres Durchgangs an beiden Sonden wurde registriert. Das markierte Gießstrangstück wurde mittels Flamme herausgeschnitten und in das Labor nach Dobrá transportiert.

Zuerst wurde hier eine nochmalige Messung der Gesamtimpulsrate mit einem Szintillationsdetektor in direktem Kontakt mit der Strangoberfläche unter Anwendung eines Bleikollimators diskontinuierlich in Abständen von 10 cm durchgeführt. An Stellen mit erhöhter Aktivität wurden durch Längs- und Querschnitte Proben entnommen. Die erhaltenen Schnittflächen wurden mit einem 35 × 35 mm-Raster versehen und an den Linienschnittpunkten ausgemessen.

Die ausgewählten Schnittflächen, besonders die aus dem Bereich der erwarteten Sumpfspitze, wurden autoradiographisch untersucht. Unter Anwendung des Filmmaterials Agfa-Gevaert STRUCTURIC D 7 P dauerte die Exposition 40 bis 120 h. Die Grenze zwischen der flüssigen und festen Phase, die auf dem Autoradiogramm durch unterschiedliche Schwärzung des mittleren Axialbereiches im Gießstrang dargestellt wird, ermöglichte die Ermittlung der Form und Masse der flüssigen Phase. Aus den gewonnenen Daten wurde die Entfernung der Sumpfspitze von der Lage des ersten Punktstrahlers im Gießstrang graphisch und mathematisch bestimmt. Dieser Wert zeigt den Abstand zwischen dem Platz, bis wohin das Blei gelangte, und dem, wo der erste Punktstrahler beim Absinken festgehalten wurde. Er diente als Korrekturfaktor zur Erhöhung der Genauigkeit der Tiefebestimmung der flüssigen Phase.

Die Identifikation verschiedener Strahler und die Bestimmung ihrer Konzentrationen in den aus dem Kristallisator gezogenen Stahlproben, in Proben aus Oberflächenzunder und in Stahlspänen, die von verschiedenen Stellen an Quer- und Längsschnitten des herausgeschnittenen Strangstücks entnommen wurden, erfolgte mittels eines Ge(Li)-Halbleiterdetektors in Verbindung mit einem Mehrkanalanalysator Typ TRACOR TN 4500. Die Proben aus dem Kristallisator sollten eine Antwort auf die Frage nach der Wirkung der Schutzhülsen und des Radionuklidübergangs in den flüssigen Stahl im Falle der sofortigen Auflösung nach dem Einwerfen geben. Eine ähnliche Information wurde auch aus den Meßergebnissen an den Zunderproben erwartet.

Auf die beschriebene Weise wurden zwei Betriebsversuche unter normalen Produktionsbedingungen beim Gießen von SM-Stahl der Qualität GOST 10 für die Rohrproduktion durchgeführt. Die Stahltemperatur in der Zwischenpfanne beim Versuch 1 betrug 1548 °C, beim Versuch 2 1537 °C, die Ziehgeschwindigkeit beim Versuch 1 0,75 m/min, beim Versuch 2 0,77 m/min.

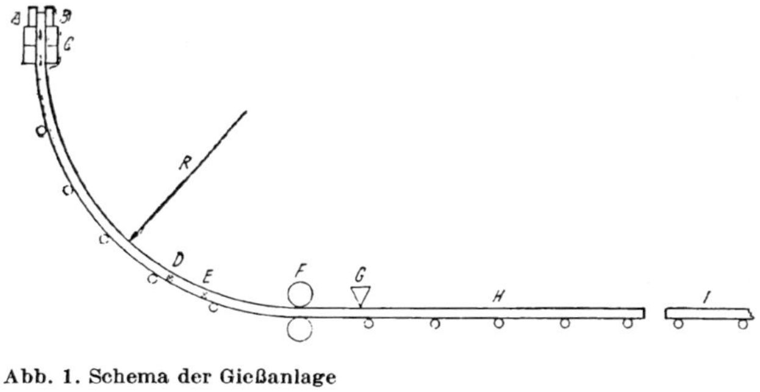

Abb. 1. Schema der Gießanlage

A – Kristallisator
B – Stahlspiegel
C – Sekundärkühlung
D – Meßstelle – Detektor 1
E – Meßstelle – Detektor 2
F – Treibrollen
G – Brennschneideinrichtung
H – Rollgang – Auslauf
I – Kühlbett
R – Gießstrangradius

Abstände B – D = 10,80 m
B – E = 11,55 m
B – G = 19,55 m
R = 9,00 m

Tab. 2. Zeitverlauf und Art der Markierung

Versuch 1				Versuch 2			
Zugabezeit		Indikator	Kapselwerkstoff	Zugabezeit		Indikator	Kapselwerkstoff
min	s			min	s		
0		–	–	0		^{141}Ce	–
1	07	^{181}Hf	Pb/Mo*)	0	55	^{110m}Ag	Mo
1	57	^{60}Co	Mo	1	30	^{60}Co	Pb/Mo*)
2	32	^{182}Ta	Stahl	2	07	^{182}Ta	Mo
3	05	^{46}Sc	Mo	2	42	^{46}Sc	Stahl
3	42	^{110m}Ag	Stahl	3	15	^{60}Co	Stahl
4	14	^{60}Co	Stahl	3	45	^{181}Hf	Mo*)
4	47	^{181}Hf	Mo*)				

*) andere Kapselgröße

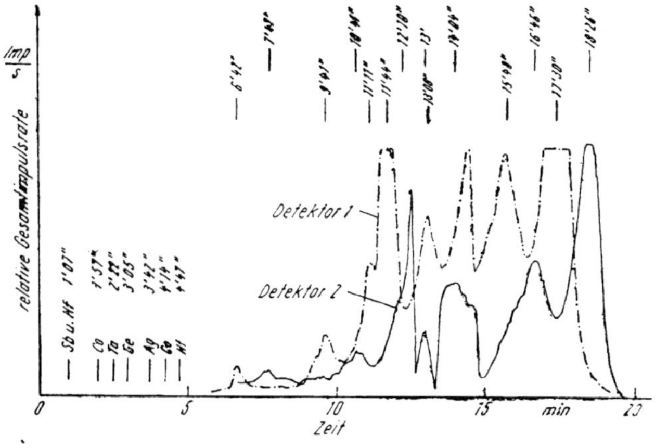

Abb. 2. Zugabe der Radionuklide und Impulsratenverlauf an den Meßstellen 1 und 2 beim Versuch 1

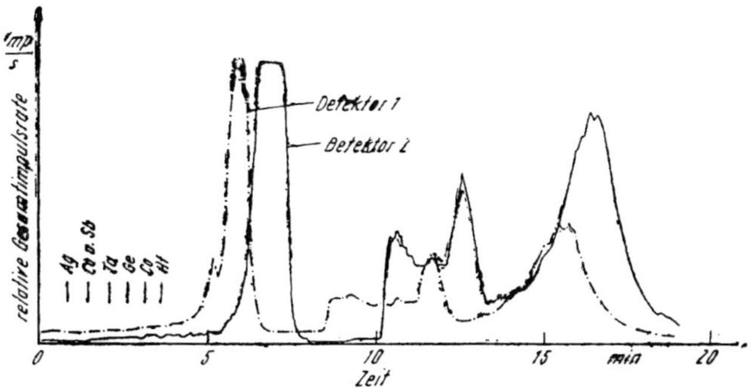

Abb. 3. Zugabe der Radionuklide und Impulsratenverlauf an den Meßstellen 1 und 2 beim Versuch 2

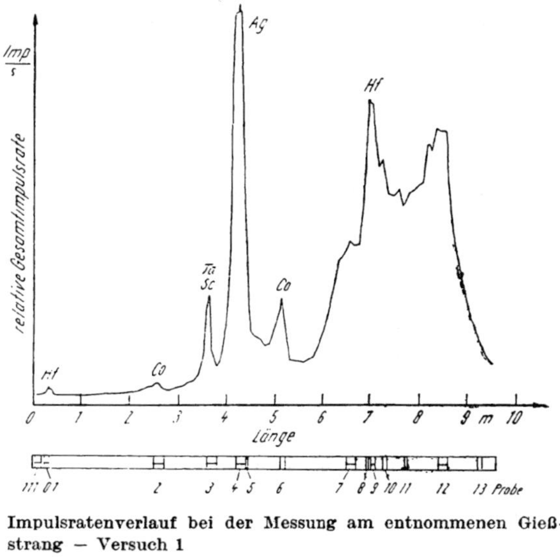

Abb. 4. Impulsratenverlauf bei der Messung am entnommenen Gießstrang — Versuch 1

Die Reihenfolge und Art der verwendeten Schutzhülsen und Indikatoren bei der Markierung des Stahls ist aus Tab. 2 ersichtlich. Den Zeitverlauf der Indikatorzugabe und der relativen Impulsraten an den Standorten der beiden Detektoren zeigt die Abb. 2 für den Versuch 1 und die Abb. 3 für den Versuch 2. Diese Diagramme wurden mittels Geräteschreiber erhalten. Die Lagen, in denen die einzelnen Radionuklide im Gießstrang festgehalten wurden, sieht man auf den Abb. 4 und 5 besser, in denen die an der Oberfläche des entnommenen Gießstrangstückes gemessenen Gesamtimpulsraten graphisch dargestellt sind.

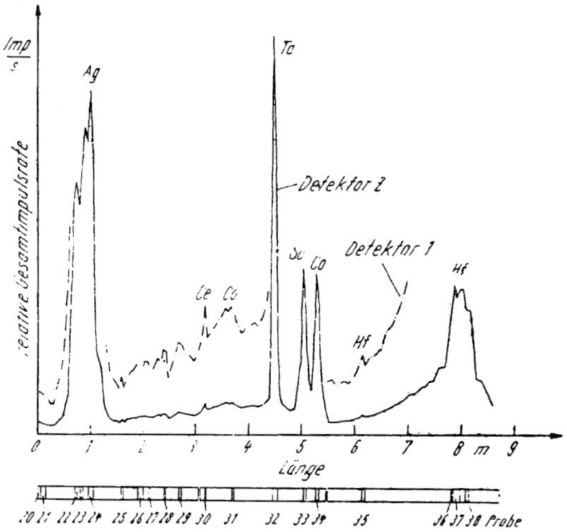

Abb. 5. Impulsratenverlauf bei der Messung am entnommenen Gießstrang — Versuch 2

Die numerierten Bereiche längs dieses Stückes zeigen die Lage und Bezeichnung der herausgeschnittenen Proben.

Die Analyse der Gamma-Spektren von Stahlspänen, die von den bezifferten Quer- und Längsschnitten des Gießstrangstückes gewonnen wurden, gab gute Informationen über die Verteilung der verwendeten Strahler im Gießstrang. Beispielsweise wurde im Versuch 1 festgestellt, daß die flüssige Phase in der Sumpfspitze durch die Radionuklide ^{181}Hf und ^{124}Sb homogen kontaminiert wurde. Beim Versuch 2 überwog dort das Radionuklid ^{110m}Ag. Einige deutliche Aktivitätsmaxima auf den Abb. 4 und 5 rühren nicht von einem, sondern von mehreren Strahlern her. Das weist auf verschiedene Auflösungs- und Bewegungsgeschwindigkeiten der Schutzhülsen und der Radionuklide im flüssigen Stahl hin.

Tab. 3. Ergebnisse der mathematischen Tiefenbestimmung der flüssigen Phase

	Versuch 1	Versuch 2
1. Entfernung vom Stahlspiegel im Kristallisator bis zum		
Detektor 1 in cm	1080	1080
Detektor 2 in cm	1155	1155
2. Ziehgeschwindigkeit des Gießstranges in cm/s	1,25	1,28
3. Theoretische Zeitdauer der Indikatorbewegung von der Zugabe in den Kristallisator bis zur Ankunft am		
Detektor 1 in s	864	842
Detektor 2 in s	924	900
4. Gemessene Zeitdauer der Indikatorbewegung von der Zugabe in den Kristallisator bis zur Ankunft am		
Detektor 1 in s	335	300
Detektor 2 in s	400	355
5. Zeit des Festhaltens des Indikators mit dem		
Detektor 1, gemessen in s	529	542
Detektor 2, gemessen in s	524	545
6. Tiefe des festgehaltenen Indikators mit dem		
Detektor 1, gemessen in cm	661	695
Detektor 2, gemessen in cm	655	699
7. Absinktiefe des Bleis (nach den Autoradiogrammen) in cm	48	74
8. Gesamttiefe der flüssigen Phase mit dem		
Detektor 1, gemessen in cm	709	769
Detektor 2, gemessen in cm	703	773

Beim Versuch 2 wurde zum Schutz des Stahlspiegels im Kristallisator ein mit dem Radionuklid ^{141}Ce in Form von CeO_2 markiertes Schutzpulver angewendet. Seine Anwesenheit wurde nicht nur im Oberflächenzunder, sondern auch in der Achse des Gießstranges festgestellt. Eine solche Schutzschicht muß man daher als eine Quelle für nichtmetallische Einschlüsse im Stahl bewerten.

Die Hauptaufmerksamkeit war der Bestimmung der Tiefe der flüssigen Phase im Gießstrang gewidmet. Eine Zusammenstellung der einzelnen Meßdaten gibt die Tab. 3 wieder. Basis der Tiefenbestimmung war die Zeitdifferenz zwischen der Zugabe der Punktstrahler in den Kristallisator und ihrer Ankunft an den Detektoren unter Berücksichtigung der Ziehgeschwindigkeit des Stranges. Zu derjenigen Tiefe, in der der erste Punktstrahler im Strang festgehalten wurde, ist der Abstand zwischen der Lage dieses Indikators auf der Längsachse des Stranges und der Sumpfspitze addiert worden. Dieser Abstand wurde aus den Autoradiogrammen entnommen. Die Auswertung beider Betriebsversuche führte zu grundsätzlich gleichen Erkenntnissen.

Nach unseren Erfahrungen kann für die praktische Tiefenbestimmung der flüssigen Phase sowohl beim vertikalen als auch radialen Strangguß die gleiche Arbeitsweise angewendet werden: Zugabe eines Punktstrahlers mit hohem Schmelzpunkt in einer doppelten Schutzhülse aus inaktivem Molybdän sowie radioaktiv markiertem Blei in den Kristallisator, Bestimmung der Tiefe, in der dieser festgehalten wird, mittels Szintillationsdetektor und der Tiefe der Sumpfspitze autoradiographisch. Für den Fall, daß ein Vielkanal-Gammaspektrometer zur Verfügung steht, kann die Anwendung weiterer Strahler für Kontrollzwecke oder zum Studium anderer metallurgischer Fragen empfohlen werden.

Eingegangen am 16. 12. 1985

Isotopenpraxis 22 (1986) 11, S. 395 bis 399

Die Kinetik der Verteilung von ^{131}I nach einer akuten pansomatischen Exposition von 390 mC · kg^{-1}

N. Stollárová, L. Mráz

(Lehrstuhl für allgemeine Biologie der Naturwissenschaftlichen Fakultät, „P. J. Šafárik"-Universität, Košice)[1])

Es wurde die Verteilung von ^{131}I 0,5 bis 24 Stunden nach einer einmaligen pansomatischen Exposition mit ionisierender Strahlung von 390 mC · kg^{-1} untersucht, wobei das Radioiod 2 und 18 Stunden nach der Ganzkörperbestrahlung verabreicht wurde. Nach der Exposition mit ionisierenden Strahlen sind zeitliche und quantitative Veränderungen der intrathyroidalen und extrathyroidalen Verteilung von Radioiod zu beobachten, was auf eine morphologische und funktionelle Depression der Organe zurückzuführen ist, womit auch eine beschleunigte Eliminierung des verabreichten Radionuklids aus dem Organismus zusammenhängt.

The study involves the tracing of iodine-131 distribution from 0·5 to 24 hours after a pansomatic exposure with an ionizing radiation of 390 mC · kg^{-1}. The iodine-131 was administrated at 2 and at 18 hours after the irradiation exposure and was followed in different organs of mice as test animals. The ionizing radiation leads to a temporal and quantitative change of iodine-131 in the intra- and extrathyroidal distribution which is caused by morphological and functional depression of the different organs. This is in turn connected with an accelerated elimination of the radioiodine from the organism.

Keywords

biological half-life; biosphere; body burden; excretion; iodine 131; irradiation; mice; radiation doses; TSH; X-radiation

Ionisierende Strahlen begleiten die Biosphäre und das sich in ihr abwickelnde Leben seit dem Zeitpunkt, zu welchem sich das Leben in seinen einfachsten Formen zu entwickeln begann. Der natürliche Untergrund an ionisierender Strahlung ist über lange Zeiträume relativ stabil. Erst in unserem Jahrhundert kam es infolge des Aufbaus großer Kernreaktoren zu einer erhöhten Kontaminierung der Biosphäre durch künstliche Radionuklide, wodurch Radioaktivität in die Umwelt gelangen kann. Aus dem gesamten Spektrum der freigesetzten Radionuklide sind ^{89}Sr, ^{90}Sr, ^{137}Cs und ^{131}I den lebenden Organismen am gefährlichsten. Aufgrund dieser Tatsache und wegen der Migration radioaktiver Stoffe in der Biosphäre ist das Studium des Metabolismus dieser Stoffe in lebenden Organismen begründet.

Da aus der Literatur nur für einige ausgewählte Organe Angaben über den Stoffwechsel von radioaktivem Iod bekannt sind, haben wir uns das Ziel gesetzt, die intrathyroidale und extrathyroidale Kinetik von Radioiod komplex im gesamten Organismus nach vorangegangener Bestrahlung zu untersuchen.

Material und Methodik

Im Experiment kamen männliche Mäuse des Stammes AB — Jena/Gat. mit einer durchschnittlichen Masse von 18–21 g zur Untersuchung. Gefüttert wurden die Tiere *ad libitum* mit pelettierter Standarddiät und Wasser. Nach einer 24stündigen Hungerperiode und darauffolgen-

[1]) Anschrift: ČS-041 67 Košice, Šrobárova 57

der Behandlung mit Lugollösung (Iod-Iodkalium-Lösung) wurden die Tiere einer einmaligen Ganzkörper-Röntgenbestrahlung mit einer Dosis von 390 mC · kg^{-1} unter nachstehenden Bedingungen ausgesetzt: Therapeutisches Röntgengerät TUR-250, 180 kV, 15 mA, Filter 0,5 mm Cu +0,5 mm Al, OK — 300 mm, Minutenleistung 52 mC · kg^{-1}. Zwei Stunden (Gruppe T_2) oder 18 Stunden (Gruppe T_{18}) nach der Exposition wurde intraperitoneal ^{131}I in Form von Na^{131}I mit einer Aktivität von 37 kBq verabreicht. Die Verteilung des Radioiods wurde 0,5, 1, 2, 3, 6, 12 und 24 Stunden nach der Applikation in den nachstehenden Proben und Organen ermittelt: Blut, Herz, Thymus, Lungen, Schilddrüse, Pankreas, Milz, Magen, Dünn- und Dickdarm, Nebennieren, Nieren, Harnblase samt Inhalt und auch das Streumaterial, welche zusammen eine Ausscheidungsprobe darstellten, und schließlich die Geschlechtsorgane und der Inhalt des Verdauungstraktes. Als Kontrolle dienten unbestrahlte Tiere, bei welchen die Distribution des Radioiods unter den gleichen Versuchsbedingungen wie bei den Versuchstieren verfolgt wurde. Die radioaktiven Proben wurden im NaI/Tl Bohrlochkristall auf dem Gerät NZQ-615 gemessen.

Die Radioiodmenge in den einzelnen Proben und Organen ist in Impulsen pro Gramm Masse und die Ausscheidung in Prozenten ausgedrückt.

Ergebnisse und Diskussion

Aus den Ergebnissen des Experiments geht hervor, daß der Radioiodspiegel in den einzelnen Proben und Organen der Tiere, die einer einmaligen Ganzkörperbestrahlung ausgesetzt wurden, in fast allen Fällen niedrigere Werte aufweist als bei der Kontrollgruppe. Eine Ausnahme bildet die Schilddrüse (Abb. 1). Die Verteilung des Radioiods in der Schilddrüse kann bei allen Gruppen (2 Versuchs- und 1 Kontrollgruppe) mit einer bimodalen Verteilungskurve charakterisiert werden, wobei bei der Kontrollgruppe der erste Anstieg der Radioiodmenge in der 2. Stunde und der zweite Anstieg in der 24. Stunde zu verzeichnen war. In der Gruppe T_2 tritt der erste Anstieg in der 1. und der zweite in der 12. Stunde ein, wobei die Radioiodmenge über die gesamte Zeitspanne von 24 Stunden im Vergleich mit der Kontrolle niedrigere Werte aufweist. In der Gruppe T_{18} erscheint der erste Anstieg in der 3. und der zweite in der 12. Stunde, wobei beide Werte des Radioiodspiegels gegenüber der Kontrollgruppe statistisch signifikant erhöht sind. Der zweite Anstieg des Radioiodspiegels in der Schilddrüse ist wahrscheinlich auf eine Reutilisierung aus dem Rezirkulationsanteil zurückzuführen. Die von uns festgestellte Akzeleration der intrathyroidalen Kinetik des Radioiods nach der Exposition mit ionisierenden Strahlen wird wahrscheinlich durch eine Erhöhung der Synthese und Sekretion des Thyreostimulationshormons (TSH) aus der Hypophyse verursacht. Das TSH übt eine aktivierende Wirkung auf das Abfangen und Oxidieren von Iod aus. Es kommt zu einer erhöhten Synthese und Sekretion von Thyroxin und zu einer schnelleren Hydrolyse von Thyreoglobulin, aus welchem Monoiodthyrosin und Diiodthyrosin freigesetzt werden. Die Deiodierung dieser iodierten Derivate führt zu einer Reutilisierung des Iods, die wir in unserem Versuch als den zweiten Anstieg der Radioiodmenge in der Schilddrüse, verursacht durch Resorption aus dem Rezirkulationsanteil, diskutieren wollen. Ein erhöhtes Auffangen von Radioiod durch die Schilddrüse nach Bestrahlung konnte von [1], [2], [3] festgestellt werden. Die genannten Autoren führen aus, daß die Aktivität der Schilddrüse im ersten Stadium der Strahlenschädigung prägnant ansteigt, was auch mit unseren Ergebnissen übereinstimmt.

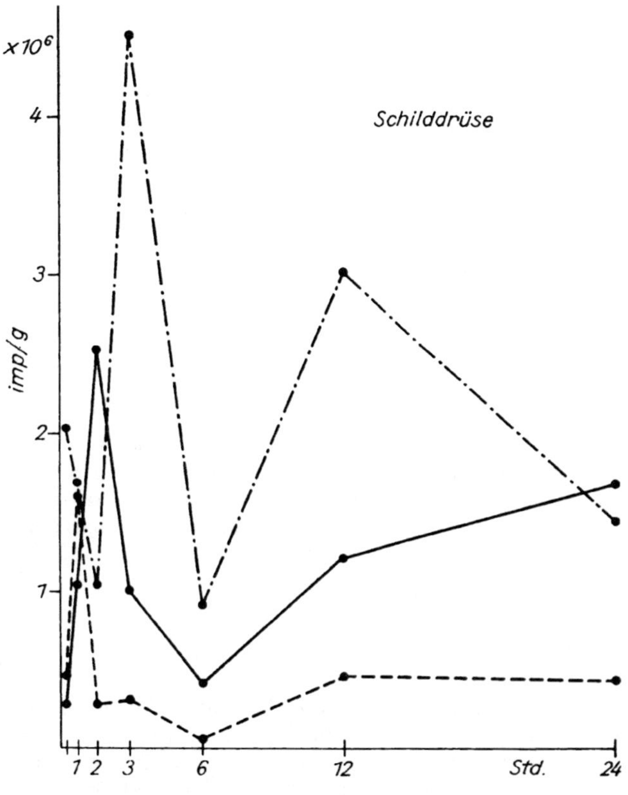

Abb. 1. Verteilung von ^{131}I in der Schilddrüse
——— Kontrollgruppe, – – – Gruppe T_2, –·–·–· Gruppe T_{18}

Im Blut (Abb. 2) sind bei den Gruppen T_2 und T_{18} im Vergleich mit der Kontrollgruppe signifikant niedrigere Radioiodspiegel zu beobachten. Dies ist wahrscheinlich auf die erhöhte Aktivität der Schilddrüse und die teilweise Störung der Bindung des Radioiods an die Globulin- und Aluminkomponente des Bluts nach der Exposition mit ionisierenden Strahlen zurückzuführen.

Auch in den Organen des Gastrointestinaltrakts ist die Radioiodmenge der Durchblutung und dem Grad des Abfangens durch die Drüsen und das Lymphsystem des Magen-Darmkanals proportional. Die Radioiodmenge ist um die durch die Schleimhaut des Magen-Darmkanals aus dem Rezirkulationsanteil resorbierte Menge erhöht. Im Magen, im Dünn- und Dickdarm, in der Leber, im Pankreas (Abb. 3) weist der Radioiodspiegel bei der Kontrollgruppe höhere Werte als bei den Versuchsgruppen auf — mit Ausnahme der 1. Stunde im Dünndam bei der Gruppe T_2. Die Magenschleimhaut resorbiert Radioiod unter physiologischen Bedingungen und besitzt die Fähigkeit dieses wieder in das Lumen auszuscheiden. Diese Fähigkeit büßt sie jedoch — wie aus den Ergebnissen ersichtlich ist — nach einer Exposition von 390 mC · kg^{-1} ein. Im Dünndarm kommt es infolge der Bestrahlung zu Epithelschäden, zu einer Abtragung der Dünndarmschleimhaut, zur Zerstörung der Peyerschen Plaques (Noduli lymphatici aggregati), zu einer Erhöhung der Permeabilität der Intestinalwand und somit auch zu einer Störung der Darm-

Mitteilungen / Information

IAEA: SICHERUNGSABKOMMEN MIT ALBANIEN

Am 1. Juli 1986 wurde zwischen der Sozialistischen Volksrepublik Albanien und der IAEA ein Sicherungsabkommen unterzeichnet, durch das alle Taetigkeiten Albaniens auf dem Nuklearsektor erfasst sind. Das Abkommen wurde vom Botschafter Albaniens, Idriz Bardhi, und vom Generaldirektor der IAEA, Dr. Hans Blix, unterzeichnet. Es ist das erste Mal, dass ein Mitgliedsstaat die IAEA um den Abschluss eines Sicherungsabkommens ersucht hat, mit dem alle Taetigkeiten des Staates auf dem Nuklearsektor erfasst werden, ohne durch den Vertrag ueber die Nichtverbreitung von Kernwaffen oder den Vertrag ueber das Verbot von Kernwaffen in Lateinamerika dazu verpflichtet zu sein. Die IAEA hat dieses Abkommen in Uebereinstimmung mit Artikel III.A.5. ihrer Satzung abgeschlossen. Durch diesen Artikel wird die IAEA ermaechtigt, auf Ersuchen eines Mitgliedsstaates Sicherungsmassnahmen auf jedwede Taetigkeit dieses Staates auf dem Gebiet der Nuklearenergie anzuwenden. Das Abkommen wurde vom Gouverneursrat der IAEA am 11. Juni 1986 angenommen. (IAEA-Pressekommunique (1986)15)

Kurznachrichten / Newlets

Strahlenschutz:
NATUERLICHE UND ZIVILISATORISCHE STRAHLENEXPOSITION

Nach Angaben des deutsch-schweizerischen Fachverbandes fuer Strahlenschutz erreicht die Strahlenexposition der Bevoelkerung in der BRD durch natuerliche und zivilisatorisch bedingte Strahlung annaehernd 4 Millisievert pro Jahr(mSv/a) bzw. in der immer noch gebraeuchlichen alten Einheit 400 Millirem pro Jahr (mrem/a). Die terrestrische Strahlung und die durch Radon in Wohnhaeusern sich ergebende Strahlenbelastung weisen regional und lokal grosse Unterschiede auf.

		mSv/a	mrem/a
Von aussen			
durch kosmische Strahlen	ca.	0,3	30
durch terrestriche Strahlung	"	0,5	50
Von innen			
vorwiegend durch Radon	"	1,5	150
Durch medizinische Anwendungen	"	1,5	150
Aus kuenstlichen Quellen (Atombombenversuche, Kernkraftwerke)	"	0,1	10
zusammen	ca.	3,9	390

also nahezu 4mSv/a (400 mrem/a)
(Atomwirtschaft/Atomtechnik 31 (1986)7)

Radiation Processing:
FOOD IRRADIATION - CLEARANCE RECOMMENDED

The Advisory Committee on irradiated and novel foods, chaired by Sir Arnold Burgen, has presented its report on the safety and wholesomeness of irradiated foods.

The report advises that:

. ionising radiation up to an overall average dose of 10 kilogray, correctly applied, provides an efficacious food preservation treatment which will not lead to a significant change in the natural radioactivity of the food or prejudice the safety and wholesomeness of the food;

. that there is no justification on public health grounds for the process not to be permitted within the dose range advised; and

. that the benefits offered by the process present strong grounds for a general clearance to be granted to this extent.

The food advisory committee recommends that irradiated foods and food containing irradiated ingredients should be clearly indicated as such throughout manufacture and at the point of sale. The advisory committee's report endorses the conclusion reached in 1980 by a joint FAO/IAEA/WHO expert committee on the wholesomeness of irradiated foods (JECFI) that the irradiation of food up to an overall average dose of 10 kilogray presents no toxicological hazard and introduces no special nutritional or microbiological problems. The JECFI's conclusion was used as the basis for the current general standard for irradiated foods which was adopted by the Codex Alimentarius Commission in 1983. The Codex Alimentarius Commission was established by the Food and Agriculture Organisation of the United Nations and the World Health Organisation to implement the joint FAO/WHO foods standards programme. The purpose of the programme is:

. to protect the health of consumers and to ensure fair practices in the food trade;

. to promote coordination of all food standards work undertaken by international governmental and non-governmental organisations;

. to determine priorities and initiate and guide the preparation of draft standards through and with the aid of appropriate organisations;

. to finalise standards and, after acceptance by governments, publish them in the Codex Alimentarius either as regional or worldwide standards.

Food irradiation (except when resulting from the use of low-dose nucleonic equipment for quality control in food processing and except to produce sterile diets for patients needing them) and the importation and marketing of irradiated foods are currently prohibited in the UK. It is permitted in other countries and is used especially for the treatment of soft fruit, fish, spices and potatoes. Ionising radiation can be used to destroy or reduce the numbers of pathogenic or spoilage organisms in food, control insect infestation of grain etc, delay the ripening of fruit, and inhibit sprouting of certain vegetables. The sources of ionising radiation are either radioactive isotopes or machines producing X-rays or electrons, and both can be used for food irradiation. The dose of radiation received by the food depends on the type of radiation source and the food's position and time of exposure in the radiation field. All these factors can be precisely controlled and recorded and, in addition, reliable methods have been developed to measure the radiation dose at various points in a food. The dose received by each food item can therefore be monitored effectively. (Atom (1986) 356)

Veröffentlichungen / Publications

UNITED KINGDOM ATOMIC ENERGY AUTHORITY

AERE-PR/EMS 11
Environmental and Medical Sciences Division progress report January-December 1983 - June, 1985, 118 pp

AERE-R 11899
A study of some actinide hydrolysis products using ultrafiltration -J.D. Wilkins; S.J. Wisbey, November, 1985, 57 pp

AERE-R 11944
The verification of NAMMU using HYDROCOIN level 1 cases 1 and 7: transient flow from a borehole and saturated flow through a shallow land disposal facility - A.W. Herbert, October, 1985, 38 pp

AERE-R 11947
Handbook of interatomic potentials. 3. Semiconductors - V.J.B. Torres; A.M. Stoneham, November, 1985, 196 pp

AERE-R 11960
Flow and transport in the unsaturated zone - C.P. Jackson, April, 1985, 7 pp

AERE-R 11975
The solubility of neptunium (I) at high pH - P.T. Ewart; S.J.M. Gore; S.J. Williams, October, 1985, 11pp

AERE-R 12010
The potential use of transmission tomographic techniques for the quality checking of cemented waste drums. Progress report to 31 March 1985 - J. Huddleston; I.G. Hutchinson, January, 1986, 22 pp

AERE-R 3506
Concentration measurements in optical particle sizing - W. Dalzell, November, 1985, 15 pp

Ausgewählte Beiträge / Recently published papers

BIOCHEM. MED. 34(1985)2
Tracer priming in human protein turnover studies with [N-15] glycine - J. Malayappa et al

BIOLOGICHESKIE HAUKI (1986) 1
Nitrate and ammonium assimilatio and the activity of nitrogen assimilation enzymes in wheat seedlings (using N-15-ammonium ions and N-15-nitrate ions) - E.D. Shirshova et al

Effect of high dispersed zink powder on cAMP phosphodiesterase activity in mices' thymocytes (using (2,8-H-3)cAMP) - S.D. Rybalkin et al

Bulg. J. Physics 13(1986)1 (in Bulgarian)
Group Theoretical Approach to Paraxial Schroedinger Equation - B.A. Nikolov

Temperature Dependence of the Positron Annihilation in BaF_2 - M.K. Georgieva et al

Positron Anihilation in γ-Irradiated LiF - M.K. Georgieva et al

JADERNA ENERGIA 32(1986)5 (in Czech)
Dosimetry of relativistic charged particle beams with solid state detectors. I. methodology - P. Spurny

Air Kerma - rate constant - J. Seda et al

Peltier battery cooled Si/Li X-ray spectrometers - V.V. Berjoskin

J. RADIOANAL. NUCL. CHEM. LETT. 104(1986)3
Extraction of technetium by crown ethers and cryptands - M.G. Jalhoom

Recent environmental tritium levels in Japan - II. Coastal seawater and lake water - N. Momoshima et al

J. RADIOANAL. NUCL. CHEM. LETT. 104(1986)4
Electrochemical oxidation of californium in carbonate solutions - V. Ya. Frenkel et al

Determination of uranium and thorium concentration in natural waters - N. Paunescu

Contribution to the radionuclide X-ray fluorescence analysis of human blood and plasma - E. Havranek et al

Application of NAA for determination of Hg, Fe, Eu, La and K in asbestos - D.K. Teherani

Reconstruction of an infinite radiation field with a complex energy spectrum - I. Petr

RADIOANAL. NUCL. CHEM. LETT.104(1986)5
Spectrophotometric study of the complexation reaction between uranium (VI) and one semicarbazone (SAS) - P.L. Lopez de Alba; E.R.Oliva

Simultaneous determination of hafnium and zirconium in low grade uranium ores using INAA - A. Mannan et al

Tc-99m-sulphur colloid: A freeze-dried, single-step kit for liver imaging - A.M. Al-Hilli et al

Measurement of uranium in human teeth and kidney stones with the fission track technique - R. Vartanian

Recovery of Cs-137 from acidic fission products solutions - I.I.L. Cunha; L.G. Andrade e Silva

Automated preparation of C-11 ethanol and C-11 butanol for human studies using positron emission tomography - G. Del Fiore et al

J. RADIOANAL. NUCL. CHEM. ART. 100(1986)1
On the characteristics and mechanism of the radiation-induced chain reaction between sulfur dioxide and molecular oxygen in acid solutions - M.B. Murathekov et al

Trace elements in the atmosphere of Ankara - M. Sabancu et al

Non-destructive analysis of inorganic impurities in Brazilian coals by epithermal neutron activation - L.F. Bellido; B.de C.Arezzo

An evaluation of the delayed neutron counting method for simultaneous analysis of uranium and thorium and for U-235/U-238 isotopic ratio determination - M.J.A. Armelin, M.B.A. Vasconcellos

Uranium fallout from the nuclear-powered satellites and volcanic eruptions - S.N. Bakhtiar et al

Determination of Th, U and Pu isotopes in atmosphere samples - F.S. Jiang et al.

Some observations on the synergistic extraction of U (VI) with double chelating agent systems - Mao jia-jun et al.

On the uso of spectrum subtraction in INAA based on short-lived radionuclides - H.A.Das

Detection limits in activation analysis using Ge(Li)-detectors installed in an underground laboratory - S. Niese; W. Helbig

Application of radiochemical separation techniques for the determination of extremely long-lived radionuclides in elemental targets - S. Theis, P. Englert

ATOM (1986) 351
Atomic energy and Harwell - some personal reminescences (Harwell's first 40 years) - L. Roberts

The fast reactor programme at Harwell - J.R. Matthews

Radioactive waste management - L.R. Fellingham

Environmental science at Harwell - G.A.Fletcher

Fuel cycle R & D - R.P. Bush

ATOM (1986) 352
Human exposure to radon decay products - A.D. Wrixon

Risk assessment for the PWR - Dr Gittus

Where will our energy come from - Dr Pooley

Pulsed plate rig aids EDRP and Thorp - P.J. Thompson

ATOM (1986) 353
Trading fund opportunities

Corrosion control in nuclear fuel reprocessing plants - D.F. Steele

Thousand-year concrete for waste repositories - A. Conway

Radiation criteria applicable to other hazards - Sir Edward Pochin

ATOM (1986) 354
Space satellites and electric propulsion systems - A.R. Martin

Fuel Technology Division's research extends endurance limits - R. Anderson

A marked expansion in nuclear power programmes in 1984 is recorded in an IAEA status report - F.J.L. Bindon

Neutral beam heating starts at JET

What is a cluster

ATOM (1986) 355
Living with Sellafield - S. Macgill

Robotic devices for nuclear plant - E. Abel

Best practicable environmental options (BPEO) a simplified version of the report prepared for the Department of the Environment by independent consultants

Environment Committee report - G. Greenhalgh

The opportunity of the 1990s - The Secretary of State for Energy's statement on nuclear power -

Earthquake surveys reviewed

ATOM (1986) 356
Nuclear waste and nimby (= not in my back yard) - Sir L. Marshall

25 years in the life of the Nestor research reactor at AEE Winfrith - S. Aindow

The first two weeks of the EDRP public inquiry

Food irradiation - Report on the Safety and Wholesomeness of Irradiated Foods

Conference Reports:
Scotland's energy future

Prototype fast reactor

ATOMWIRTSCHAFT/ATOMTECHNIK 31(1986)1
High Tech in Deutschland - Chancen und Herausforderungen - K.H. Beckurts

Schutz gegen Bethe-Tait-Stoerfaelle im SNR-300 - H. Huebel; H. Vossebrecker

Welt-Elektrizitaetsversorgung - H. Magerl

Spaltgasfreisetzung aus Brennelement-Endlagergebinden im Endlager - P. Arntzen et al

Die Chancen des Fusionsreaktors - A. Schlueter

HF-Anordnung zur Aufheizung von Plasmen im Bereich der Ionenzyklotronfrequenz - F. Wesner et al

Kernkraftwerke - Verursacher von Waldschaeden? - R. May

Stoermelde- und Messwerterfassungsanlagen am FMRB - W. Hajak et al

Erfahrungen beim Konservieren des KKW Zwentendorf - W. Binner

ATOMWIRTSCHAFT/ATOMTECHNIK 31(1986)2
Die Nichtverbreitung in der Ueberpruefung - G. Stein

Sicherheit bei der Befoerderung von radioaktiven Stoffen - K. Ridder, G. Wehner

Internationale Harmonisierung im Bereich des Kernenergierechts - H.-P. Bochmann

MZFR-Erstes Kernkraftwerk aus deutscher Entwicklung - Ein Rueckblick - G. Brudermueller

Aktivitaeten zur Neubewertung der Quellterme - H. Karwat

Behandlung von Stoerfaellen im Betriebshandbuch des Kernkraftwerks Gundremmingen - W. Reim

KKW-Fernueberwachung im funktionalen Zusammenhang - H. Wolf

ATOMWIRTSCHAFT/ATOMTECHNIK 31(1986)3
Werkstoffe fuer Leichtwasserreaktoren - E. Tenckhoff; U. Roesler

Werkstofftechnik fuer Brennelemente und Kernbauteile in wassergekuehlten Reaktoren - R. Holzer; H. Stehle

Metallische Komponenten fuer Hochtemperatur-Reaktoranlagen - H. Nickel et al

Werkstoffe fuer schnelle Brutreaktoren - W. Dietz et al

Werkstoffe fuer Fusionsreaktoren - K. Kummerer; K. Ehrlich

ATOMWIRTSCHAFT/ATOMTECHNIK 31(1986)5
Vertrauen und Zuversicht in die Kerntechnik - C. Berke

Genehmigungsverfahren nach Recht und Gesetz - R. Jochimsen

Sichere und kostenguenstige Energie ist lebenswichtig - M. Gruener

Langzeitsicherheit von Endlagern - D. Ehrlich et al

Die internationale Ueberwachung der friedlichen Nutzung der Kernenergie - P. Tempus

Die Sicherheitsueberwachung durch Euratom - W. Gmelin

Kernkraftwerke kleiner und mittlerer Leistung - R. Schmidt

ATOMWIRTSCHAFT/ATOMTECHNIK 31(1986)6
Der sowjetische Druckroehren-Siedewasserreaktor Anlagenbeschreinung und erste Unfallbetrachtung - J. Wolters et al

Der Fortgeschrittene Druckwasserreaktor (FDWK) - B. Kuczera et al

Brennelement-Fabrikation-Stand und Weiterentwicklung - K.G. Hackstein; E. Zastrow

Die Inbetriebnahme des SNR-300 - W. Buerkle

Neue Kernkraftwerke in Europa 1986 (Teil II) - atw-report

ATOMWIRTSCHAFT/ATOMTECHNIK 31(1986)7
Die Urananreicherungsanlage Gronau - H. Mohrhauer

Verfuegbarkeit der Kernkraftwerke in der Welt im Jahre 1985 - A. Szeless, W. Zimmermann

Die Kernenergie hat eine Zukunft - W. Braun

Kernkraftwerke in den USA 1985 - Statistik

Tschernobyl-Unfall: Abschaetzung der Aufnahme des abgelagerten Cs-137 und Sr-90 ueber die Wurzel - W. Steffens et al

Tschernobyl-Unfall: On-Line-Messung der spezifischen Aerosolaktivitaet - J. Adolph et al

Auswirkungen des Reaktorunfalls in Tschernobyl - H. Kiefer

ATOMKERNENERGIE/KERNTECHNIK 48(1986)2
Radiotracer in der Industrie - R. Link et al

Eine Zweikanal-Gammastrahlenmethode zur Dreikomponentenanalyse in hydraulischen Transportsystemen - H.-U. Fanger et al

On-Line-Bestimmung von Blei mit dem Roentgenfluoreszenz-Analysenverfahren GRAF in der Blei-Zink-Erzaufbereitung - A. Lubecki

Grundwassermessungen mittels radioaktiver Indikatoren - W. Drost

Verfahren zur beruehrungslosen Messung der Wandstaerke von Rohren mittels Rueckstreuung ionisierender Strahlung - E. Frevert

Neue bildgebende Systeme fuer die Durchstrahlungspruefung mit Radionukliden - R. Link et al

Mobiles Computer-Tomographie-Geraet zur Untersuchung ortsfester Objekte - A. Habermehl et al

Entwurfskriterium bei berohrten solar beheizten Hohlraum-Strahlungsempfaengern - K. Bammert; A. Hegazy

IAEA-BULLETIN 27(1985)4
Building on experience: An international perspective - M.A. Feraday

The European Community's research programme: Next phase - B. Huber

Overview of cost estimates and financing practices - Pabrita L. De; E.G. Delaney

Steady progress at TMI-2 -Cynthia J. Hess; S.W. Metzger

French experience and plans - A. Cregut

Taking Canada's Gentilly-1 to a "static state" - B. Gupta

Decommissioning US DOE nuclear facilities - E.G. Delaney

Remote-controlled equipment for decommissioning

Export financing in France - J. Richard-Mounet

Financing electric power in developing countries - A. Heron

Nuclear power programmes in developing countries: Costs and financing - J.-P. Charpentier; L. Bennett

Plant cultivation and breeding: Promise of technology - A. Micke

IAEA 28th General Conference: Session highlights

IAEA BULLETIN 28(1986)1
World status of radioactive waste management - D.E. Saire

Overview of IAEA activities in the field - J. Hirling et al

Long-term storage and disposal of spent fuel - A. Nechaev et al

Assessing the radiological safety of waste repositories - G. Linsley

Update: The international Stripa project - H. Carlsson

Deep-sea disposal: Scientific bases to control pollution - A. Hagen; B. Ruegger

Report from Monaco: Waste management and the sea - R. Fukai

Policy and practice in India - N.S. Sunder R Rajan

Sweden: Policy and licensing - A. Larsson et al

French experience and plans - J.J. Lefevre

Managing high-level waste in the USA- B.C. Rusche

ATOMNAJA ENERGIJA 61(1986)3
Stationary penetrability of non-equilibrium ions and atoms of gas through a partition - A.A. Pisarev; V.M. Smirnov

On calculating the sediment section of a mixed sediment extractor - S.M. Karpacheva

Joint determination of concentratrion of Rn-222 and Rn-220 decay products in air - M.V. Terentyev

Pu migration in soils - F.I. Pavlotskaya et al

The measurement of alpha values at Pu-239 resonances - Yu.V. Adamchuk et al

Investigation of the interaction of very cold neutrons with Uranium and its isotope U-235 - A.V. Antonov et al

Determination of Li concentration in silicon by a radioactivation method on a cyclotron - S. Mukhammedov et al

Determination of the yields of delayed neutrons from heavy-nuclei fission products by means of semi-empirical formulas - A.I. Lendel et al

A technical-economic evaluation of radiation shielding of industrial electron accelerators - V.V. Krayushkin et al

On the possibility of obtaining quasi-monoenergetic X-rays in the medium-energy range by means of Barkla polarizers - I.A.Tolokonnikov

KfK-NACHRICHTEN 18(1986)1
Development of New Remote Handling Systems for Hot Cells - G. Boehme

Development of High Flexible Handling Devices - W. Koehler et al

TFTR Maintenance Manipulator - M. Selig

CAD with Applicon Bravo! - Two Years of Practice at the IRE - F. Katz et al

Improvements of Servo Manipulators by Digital Control - H. Becker et al

Toxicological and Radiobiological Investigations using Tissue Cultures: Substitute and Supplement for Animal Experiments - H. Dertinger et al

Problems with the analysis of polychlorodibenzodioxins and -furans in fly ash of municipal incinerators - L. Stieglitz et al

DOKLADI AKADEMII NAUK 287(1986)2 (in Russian)
K-Ar age of the alkaline rocks of west Kamschatka - Kh.I. Amirkhanov et al

DOKLADI AKADEMII NAUK 287(1986)4 (in Russian)
K radiogeochemical specialization of metasomatites from a Bi deposit - A.M. Portuov; S.V. Ezhov

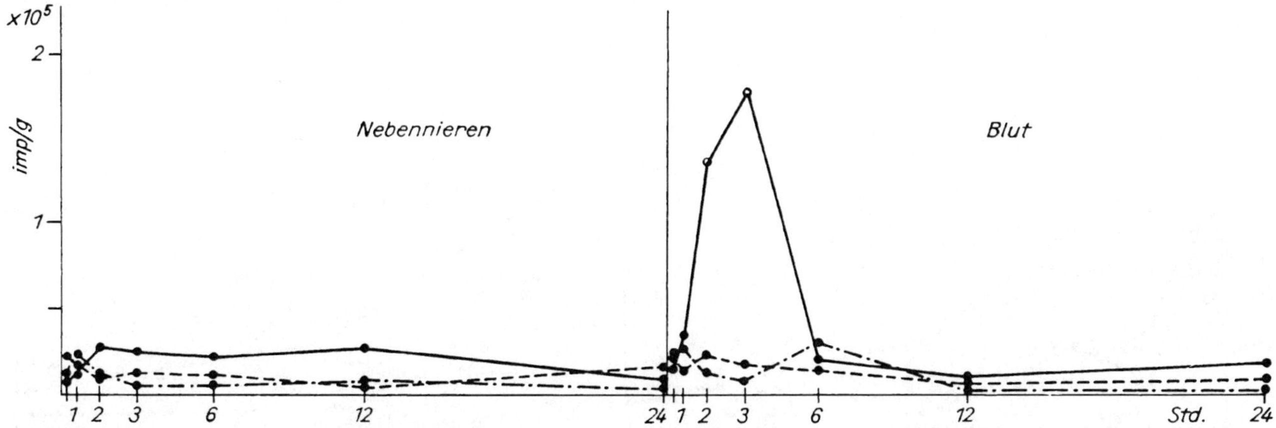

Abb. 2. Verteilung von ^{131}I in den Nebennieren und im Blut ——— Kontrollgruppe, – – – Gruppe T_2, –·–·– Gruppe T_{18}

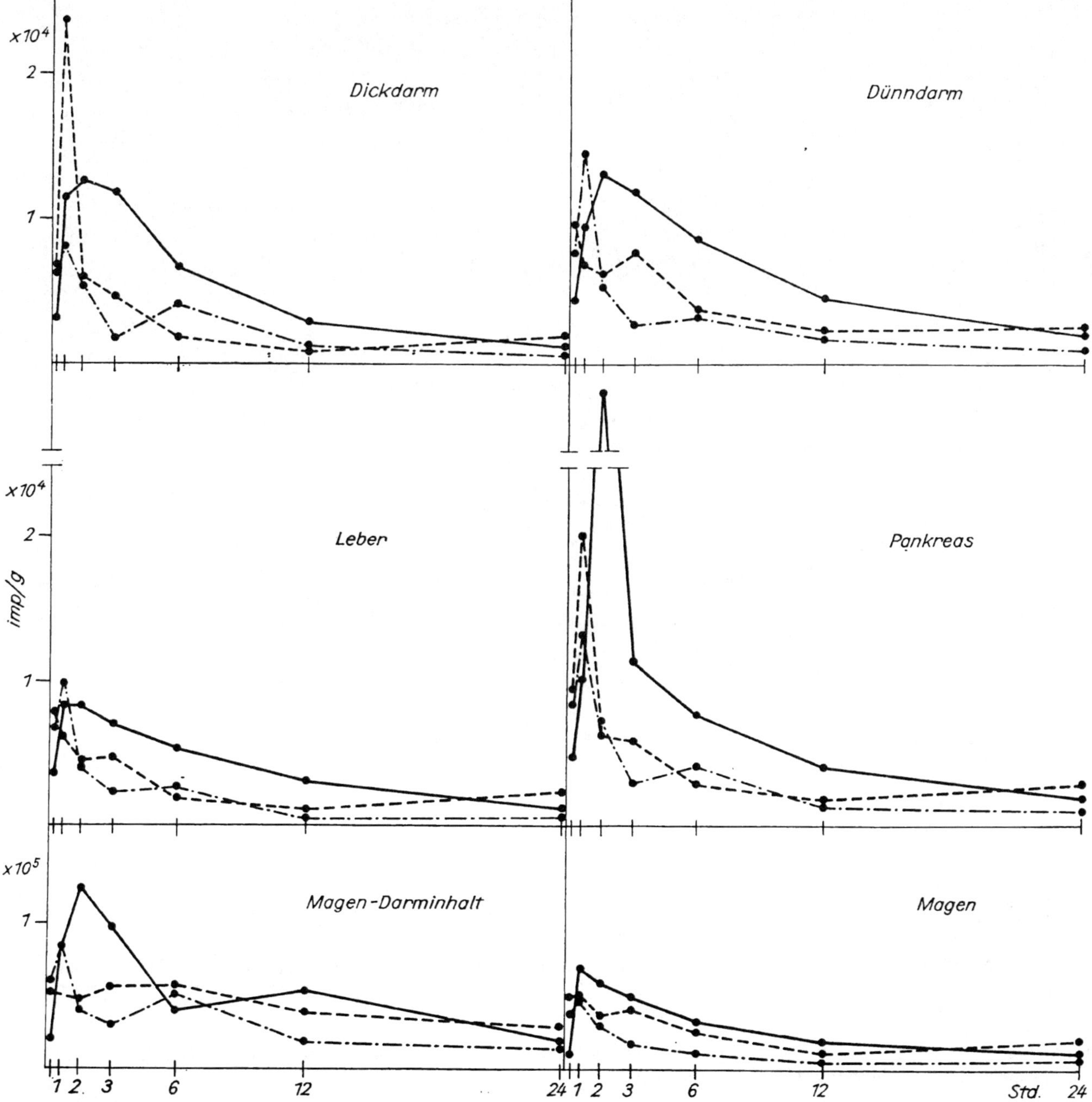

Abb. 3. Verteilung von ^{131}I im Dünn- und Dickdarm, der Leber, dem Pankreas, im Inhalt des Gastrointestinaltraktes und im Magen ——— Kontrollgruppe, – – – Gruppe T_2, –·–·– Gruppe T_{18}

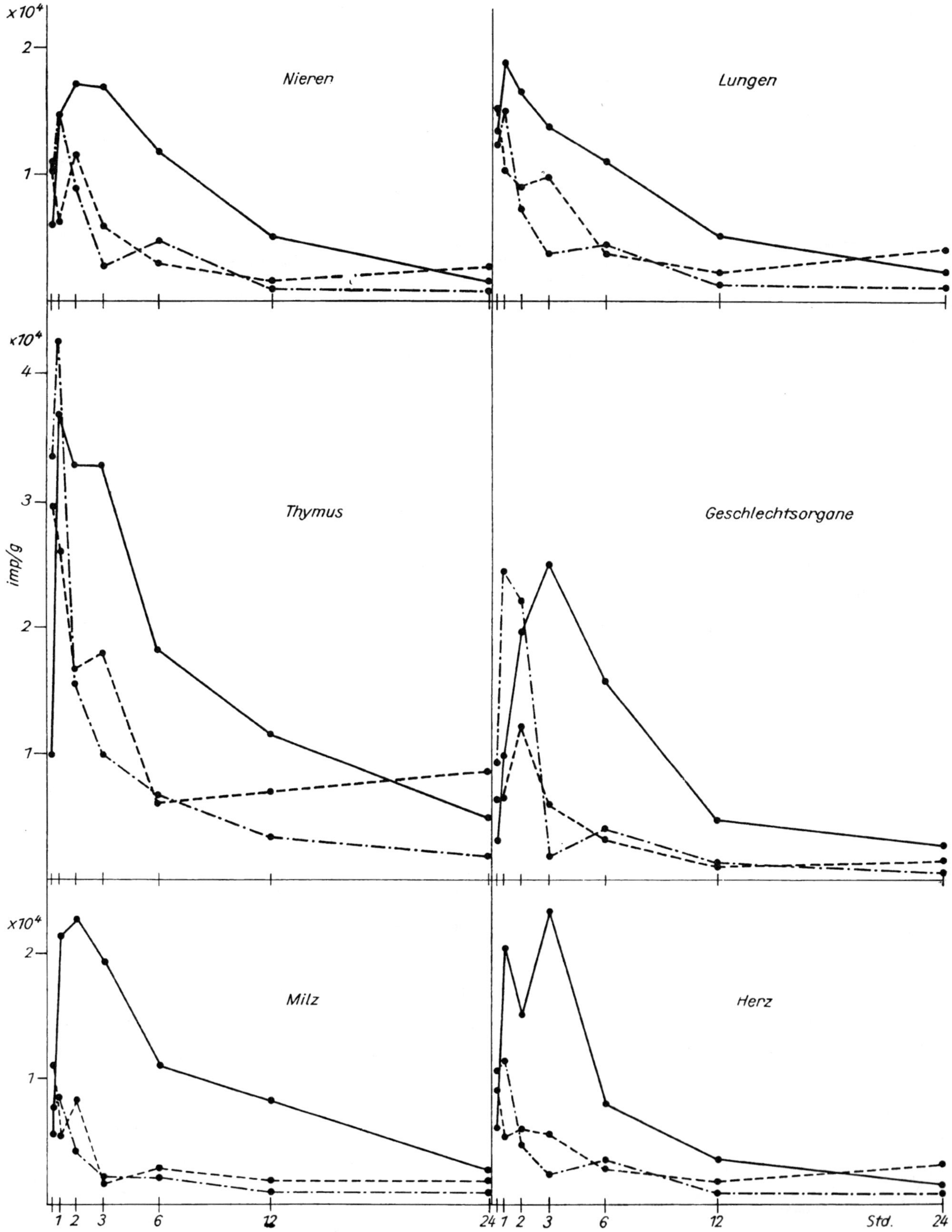

Abb. 4. Verteilung von ^{131}I in den Nieren, den Lungen, im Thymus, den Geschlechtsorganen, in der Milz und im Herzen
—— Kontrollgruppe, – – – Gruppe T_2, –·–·– Gruppe T_{18}

barriere, was ab der 2. Stunde einen niedrigeren Radioiodspiegel im Dünndarm zur Folge hat. Die niedrigeren Werte des Radioiodspiegels im Dickdarm sind auf Veränderungen der Motilität und Permeabilität des gesamten gastrointestinalen Trakts nach der Bestrahlung zurückzuführen [4].

In den Nieren (Abb. 4), den Lungen (Abb. 4), den Geschlechtsorganen (Abb. 4) und im Herz (Abb. 4) ist eine deutlich veränderte Permeabilität der Zellmembranen zu verzeichnen, woraus sich niedrigere Werte des Radioiodspiegels gegenüber den Kontrolltieren im gesamten verfolgten Intervall ergeben. Die Strahlenempfindlichkeit des Lungengewebes ist auf die Durchblutung des Lungenparenchyms zurückzuführen, was einen verstärkenden Sauerstoffeffekt zur Folge hat [5]. Obwohl der Herzmuskel zu den weniger strahlenempfindlichen Geweben gehört, treten auch hier strahlenbedingte funktionelle und morphologische Abweichungen auf, die meist eine Folge von Herzgefäßschäden sind [6].

Strahlenschäden am Keimepithel der Geschlechtsdrüsen kommen in einer Verminderung des Volumens sowie der Masse der Keimdrüsen zum Ausdruck, womit auch der niedrigere Radioiodspiegel in diesen Organen nach der Bestrahlung zusammenhängt [7].

Strahlenschäden am hämatopoetischen Apparat (Abb. 4) haben statistisch signifikant niedrigere Radioiodspiegel in der Milz zur Folge. Mehrere Autoren [8], [9] berichten, daß nach der Bestrahlung chemische und biochemische Veränderungen eintreten, denen morphologische Veränderungen vorangehen bzw. sie verlaufen parallel mit denselben. Bei hohen Expositionen ist eine prägnante Atrophie der Milz zu beobachten, deren Masse auf 1/3 bis 1/5 des ursprünglichen Werts sinken kann [10].

Im Thymus (Abb. 4) sind ab der 1. Stunde (Gruppe T_2) bzw. ab der 2. Stunde (Gruppe T_{18}) bis zur 24. Stunde statistisch signifikant niedrigere Radioiodspiegel als bei den Kontrollgruppen zu verzeichnen.

Die strahlenbedingten Veränderungen in den Nebennieren beeinflussen in bedeutendem Maße die Reaktion des gesamten Organismus auf die ionisierenden Strahlen. Nach höheren Expositionen und vorangegangener Stimulierung tritt eine Verringerung der adrenokortikalen Reserve ein und die Reaktion der Nebennieren auf die Belastung ist trotz der noch erhaltenen Fähigkeit der Steroidgenese deutlich vermindert [11].

Die Ausscheidung des Radioiods im Harn (Abb. 5) stellte in unserem Experiment die in der Harnblase und in der Streu festgestellte Gesamtmenge dar. Unsere Versuchsergebnisse zeugen von einer erhöhten Ausscheidung des Radioiods im Harn nach der Exposition mit ionisierenden Strahlen.

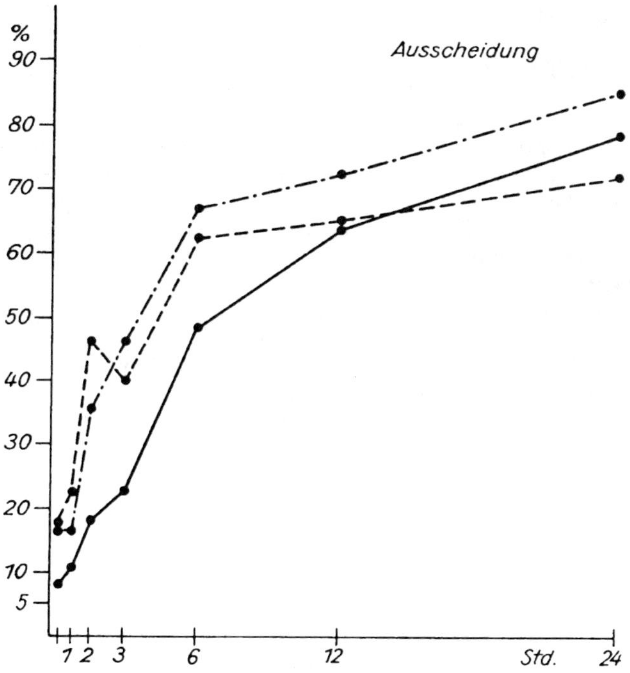

Abb. 5. Ausscheidung von ^{131}I in Prozenten

Zusammenfassung

Nach einer Exposition mit ionisierenden Strahlen von 390 mC · kg^{-1} wurden in einzelnen Organen und Proben niedrigere Werte des Radioiodspiegels und eine erhöhte Ausscheidung von Jod-131 gegenüber den Kontrolltieren gefunden. Diese Erscheinung ist auf eine gestörte Bindung des Radioiods an die Makromoleküle und auf die veränderte Permeabilität der Zellmembranen zurückzuführen. Diese Faktoren verursachen eine morphologische und funktionelle Depression der Organe und eine Veränderung des gesamten Polykomponentensystems des lebenden Organismus nach Exposition mit ionisierenden Strahlen. Der statistisch signifikant höhere Radioiodspiegel in der Schilddrüse mit einem 12stündigen Abstand der Exposition nach Beginn der Verfolgung der Verteilung ist auf eine Hypophysenstimulation durch die ionisierenden Strahlen zurückzuführen.

Eingegangen am 4. 12. 1985

Literatur

[1] *W. R. Greig*, J. Clin. Endocrinol. 25 (1963) 1411
[2] *J. I. Moskalev, G. M. Jegorova*, Biologitscheskije dejstvije vneschnych i vnutrennych istocnikov radiacii. Moskva: Medicina 1972, 355
[3] *H. Lach, B. Dymarczyk*, Acta Biol. Acad. Sci. Hung. 27 (1976) 9
[4] *Y. T. Kurcin*, Effect of Ionizing Radiation on the Digestive System. Amsterdam — London: Elsevier Publ. Co., 1963, 326
[5] *D. Teales, G. J. Cooper*, Amer. J. Röntgenol. 96 (1966) 612
[6] *C. C. Berdjis*, Fed. Proc. 27 (1968) 414
[7] *J. Horák, A. Horáková*, Folia Veter. 25 (1981) 59
[8] *M. Hill, M. Praslička*, Čs. biológia 6 (1957) 6
[9] *S. W. Jordan*, Exp. Mol. Pathol. 6 (1967) 156
[10] *N. Stollárová*, Dissertation, P. J. Šafárik-Unviersität. Košice 1982
[11] *Y. Z. Achmetov*, Zool. žurnal 2 (1973) 238

Isotopenpraxis 22 (1986) 11, S. 400 bis 407

Rechnergestützte Verarbeitung und Auswertung von NMR-Spektren in der Isotopen- und Strahlenforschung

E. Winkler, H. Sprinz, B. Lindemann, U. Hädrich, U. Franck

(Akademie der Wissenschaften der DDR, Zentralinstitut für Isotopen- und Strahlenforschung)[1])

Zur automatisierten Verarbeitung und Auswertung von NMR-Spektren bei der Lösung von Problemen der Isotopen- und Strahlenforschung wurde ein NMR-Spektrometer BS 487 C/Tesla über einen Mikrocomputer AMCA-80 mit einer EDVA EC 1040 on-line gekoppelt. Entsprechend den unterschiedlichen Aufgabengebieten wurden 3 Verarbeitungsprogramme (FORTRAN) für den Rechner EC 1040 entwickelt. Das Programm LINFOR, eingesetzt zur Analyse der Strahlenwirkung auf Membranvesikel, trennt ein NMR-Spektrum in 2 Linien bei gleichzeitiger Bestimmung der Linienformparameter. Für die Reinheitsanalytik deuterierter Verbindungen ermöglicht das Programm AKKU eine Spektrenakkumulation (max. 255 Akkumulationen). Das Programm ISOTAN ermöglicht, bei deuterierten Lösungsmitteln den Deuterierungsgrad verschiedener chemischer Gruppen zu ermitteln. Durch Beispiele wird die Nutzung der Verarbeitungsprogramme belegt.

With the aim of automated processing and interpretation of NMR spectra in the field of isotope and radiation research a BS 487 C/Tesla NMR spectrometer was linked via an AMCA-80 microcomputer with an EDPM EC 1040. In order to cover different research problems 3 programs were written in FORTRAN for the EC 1040 computer. The program LINFOR (used for the analysis of the effect of ionizing radiation on membrane vesicles) separates an NMR spectrum into 2 lines and simultaneously determines the line shape parameters. In order to check the purity of deuterated substances the program AKKU performs an accumulation of spectra (max. 255 scans). The program ISOTAN enables the degree of deuteration of different chemical groups in deuterated solvents to be determined. Examples illustrate the application of the programs.

Keywords

computer codes; data processing; deuterium compounds; nmr spectrometers; on-line systems

Einleitung

Zur Lösung von Problemen der Isotopen- und Strahlenforschung kann die NMR-Spektroskopie in vielfältiger und unentbehrlicher Weise eingesetzt werden. Im Rahmen des hier dargestellten Verfahrens zur automatisierten Verarbeitung und Auswertung von NMR-Spektren sind drei Aufgabengebiete zu nennen. Für die Reinheitsanalytik deuterierter Lösungsmittel ist ein möglichst hohes Signal-Rauschverhältnis anzustreben, um auch schwache Verunreinigungen, die bei der Anwendung der Lösungsmittel störend wirken, nachweisen zu können. Die zu Verunreinigungen gehörenden Linien sind aus dem Spektrum herauszusuchen, und deren Flächen sind im Vergleich zu der eines Referenzsignals zu bestimmen. Für deuterierte Lösungsmittel ist weiterhin der Deuterierungsgrad verschiedener chemischer Gruppen zu ermitteln, ebenfalls über eine Liniensuche und Flächenbestimmung. Schließlich ist zur Analyse der Strahlenwirkung auf Membranvesikel eine Trennung des ^{1}H-NMR-Signals in zwei Linien und eine Bestimmung der Signalparameter (Linienformanalyse) notwendig.

Zur Spektrenaufnahme dient in der hier dargelegten Gerätekonfiguration ein NMR-Spektrometer BS 487 C (Tesla, Brno). Zur Steuerung des Meßwertdurchlaufs sowie zur Übernahme und Vorverarbeitung der Meßdaten ist das Spektrometer mit einem Mikrorechner AMCA-80 on-line gekoppelt. Die Auswerteprogramme zur Lösung der drei obengenannten Probleme werden an einer EDVA EC 1040 abgearbeitet. Zur Kopplung und Datenübertragung ist das System CAMAC eingesetzt. Die Meßwerterfassung und -verarbeitung erfolgt im Dialog unter Realzeitbedingungen.

Die eingesetzte Hardware, die Programmierung des Mikrorechners sowie die Steuerung des Experiments im Dialogbetrieb wurden an anderer Stelle ausführlich vorgestellt [1]. Die vorliegende Publikation hat die Auswerteprogramme für den Rechner EC 1040 sowie den Informationsfluß zwischen der Konfiguration Spektrometer-Mikrorechner und dem Großrechner zum Inhalt.

Ablaufsteuerung und Datenfluß

Abb. 1 zeigt die Verknüpfung der drei Auswertungsprogramme mit den Hardware- und Software-Komponenten des Programmsystems zur on-line-Kopplung von Experimenten an die EDVA EC 1040.

Das System benutzt zur Datenübertragung von und zu den Experimenten in der EDVA EC 1040 die Zugriffsmethode ETOS. Dadurch wird weitgehend die Struktur des Programmsystems bestimmt, welches aus einem generierbaren Monitor und einer Reihe von Steuer- und Verarbeitungsprogrammen besteht. Zur Realisierung der on-line-Kopplung des NMR-Spektrometers sind im Programmsystem neben den Programmen zur Meßwertverarbeitung folgende Programme vorhanden: Nachrichtensteuerprogramm (NMNSP), Nachrichtenverarbeitungsprogramme (NMNVP, NVPNMR), Fehlermaßnahmeprogramm (NMFMP).

[1]) Anschrift: DDR-7050 Leipzig, Permoserstr. 15

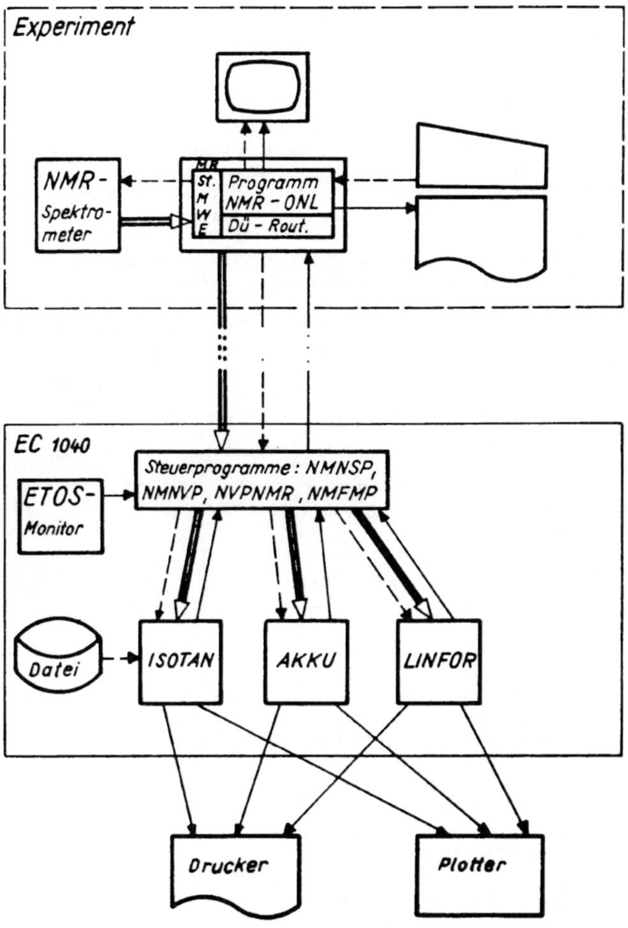

Abb. 1. Übertragung von Meßdaten, Auswerteparametern und Ergebnissen zwischen den Hardware- und Softwarekomponenten des on-line-Auswertungssystems
⟹ Meßdaten
– – → Parameter, Steuerwerte
——→ Ergebnisse

Das Nachrichtensteuerprogramm wird vom Monitor sowohl nach dem Empfang einer Nachricht vom Experiment als auch nach dem Senden einer Nachricht an das Experiment aktiviert. Empfangene Nachrichten werden vom Übertragungscode (KOI-8) in den internen Code (DKOI) übersetzt und an das Nachrichtenverarbeitungsprogramm NMNVP übergeben. Nach Abschluß des Sendens einer Folge von Nachrichten an das Experiment sorgt das Nachrichtensteuerprogramm für eine erneute Verbindungsaufnahme zum Experiment.

Treten E/A-Fehler in den Phasen der Verbindungsaufnahme, des Sendens oder des Empfangs von Nachrichten auf, so wird das Fehlermaßnahmeprogramm aktiviert. Es unterbricht die Übertragung, versetzt die Leitung in den Zustand „nicht aktiv" und informiert den Bediener des Zentralrechners. Nach eventueller Beseitigung der Fehlerursache kann der Bediener durch die Eingabe eines Kommandos eine erneute Verbindungsaufnahme zum Experiment einleiten.

Das Nachrichtenverarbeitungsprogramm NMNVP übernimmt vom Nachrichtensteuerprogramm die empfangenen Auswerteparameter und Meßdaten. Zu deren Verarbeitung wird das Programm NVPNMR aufgerufen. Dessen Aufgabe besteht vor allem in der Prüfung der Auswerteparameter und Meßdaten und deren Aufbereitung für die nachfolgenden FORTRAN-Verarbeitungsprogramme. Bei den Auswerteparametern handelt es sich um eine für jede Verarbeitungsart (Akkumulation, Linienformanalyse, Isotopenanalyse) fest definierte Anzahl von Positionsparametern sowie organisatorischen Angaben (Spektrennummer, Datum etc.), die vom Experimentator einzugeben sind. Das Programm NVPNMR prüft diese Parameter auf syntaktische Richtigkeit und Vollständigkeit. Im Fehlerfall wird eine Mitteilung an den Experimentator gesendet, der daraufhin die Eingabe wiederholen kann. Die Parameter werden vom Programm NVPNMR in INTEGER-, REAL- bzw. LOGICAL-Größen umgewandelt und in einem COMMON-Block bereitgestellt.

Meßdaten werden in einem 4 -KByte-Puffer in Form von 2 K INTEGER-Halbworten bereitgestellt. Der Puffer enthält jedesmal die digitalisierten Ordinatenwerte eines NMR-Spektrums. In Abhängigkeit von der gewünschten Verarbeitungsart, die der Experimentator durch Eingabe der Verarbeitungskennzeichen (AKZ) L (Linienformanalyse), I, X bzw. Y (Isotopenanalyse) oder A (Akkumulation) festlegt, werden die entsprechenden FORTRAN-Verarbeitungsprogramme LINFOR, ISOTAN oder AKKU aufgerufen. Nachrichten, welche im Ergebnis der Verarbeitung an den Experimentator gesendet werden sollen, werden als Folge von Sätzen fester Länge in einer sequentiellen Datei auf Magnetplatte hinterlegt. Nach Rückgabe der Steuerung an das Programm NMNVP werden diese Sätze wieder gelesen und als Nachrichtenfolge an den Experimentator gesendet. Danach wird eine erneute Verbindungsaufnahme durch NMNSP eingeleitet, so daß ein weiterer Zyklus, bestehend aus Eingabe von Auswerteparametern und Meßdaten, Verarbeitung dieser Werte und Ausgabe von Nachrichten stattfinden kann (wobei die Eingabe von Auswerteparametern entfallen kann, falls der zuletzt gesendete Datensatz gültig bleiben soll).

Die Verarbeitungsprogramme LINFOR, AKKU und ISOTAN sind in FORTRAN geschrieben und werden als SUBROUTINE-Unterprogramme vom Nachrichtenverarbeitungsprogramm NVPNMR aufgerufen. Jedes Unterprogramm enthält nur einen Eingang, Verzweigungen erfolgen (bei ISOTAN) innerhalb des Unterprogramms. Die Auswerteparameter stehen in einem COMMON-Bereich, die Meßdaten werden als aktuelle Parameter beim Unterprogrammaufruf übergeben. Der COMMON-Bereich enthält auch Variable, deren Wert sowohl im Unterprogramm als auch im aufrufenden Programm verändert wird.

Resultate werden, wie oben beschrieben, an den Experimentator über die DÜ-Leitung ausgegeben. Zusätzlich wird ein Protokoll der jeweiligen Spektrenauswertung mit dem Paralleldrucker des EC 1040 erstellt. Über den Digitalplotter DIGIGRAF ist die Ausgabe graphischer Darstellungen möglich.

Verarbeitungsprogramm LINFOR

Das Programm LINFOR wurde zur Bearbeitung einer strahlenbiologischen Aufgabe erstellt [2] und kann darüber hinaus bei allen Problemen eingesetzt werden, die ein Spektrum ergeben, welches durch Gl. (1) beschreibbar ist.

Aus dem Spektrum (Abb. 2) ist eine Reihe von Parametern für die beiden Signale zu ermitteln: Lage (Abstand) der Signale, Charakteristika für die Signalbreiten (Halbwertsbreiten), Gesamtfläche und relative Flächenanteile der Einzelsignale. Gleichzeitig ist eine Nullinienkorrektur vorzunehmen.

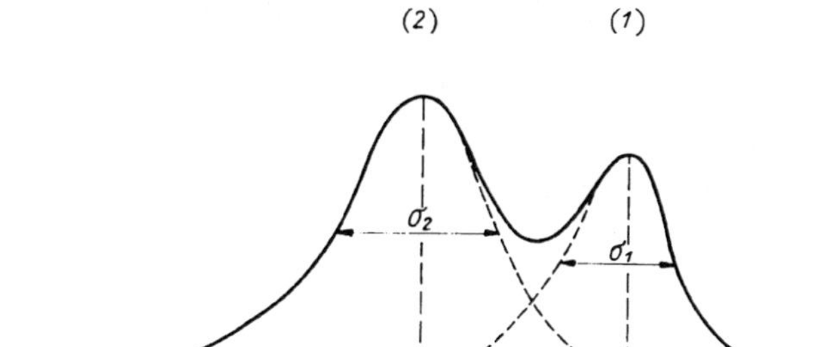

Abb. 2. Programm LINFOR: Spektrum, bestehend aus zwei Linien mit Angabe der Linienformparameter

Die Signale und die Nullinie werden durch ein mathematisches Modell beschrieben. Zur Darstellung der Signalform hat sich für das zugrundegelegte Problem eine *Lorentz*-Kurve als am geeignetsten erwiesen [3]. Für die Nullinie wird ein Orthogonalpolynom (*Legendre*-Polynom) vom Grad 2 angesetzt [4, 5], welches den Vorteil bietet, daß sich bei der Parameterschätzung jeder Term und damit jeder Koeffizient a_i durch Wegfall der Kopplungen im Normalgleichungssystem einzeln und unabhängig von allen anderen ergibt, wodurch sich die Konvergenzgeschwindigkeit des verwendeten Schätzalgorithmus erhöht.

Das gesamte Spektrum wird somit dargestellt durch

$$f(x) = C\left\{\frac{q_1}{\pi\sigma_1/2}\,\frac{1}{1+\left(\dfrac{x-\mu_2+\Delta_1}{\sigma_1/2}\right)^2} + \frac{(1-q_1)}{\pi\sigma_2/2}\,\frac{1}{1+\left(\dfrac{x-\mu_2}{\sigma_2/2}\right)^2}\right\} + a_0 + a_1x + a_2\cdot\frac{1}{2}(3x^2-1) \qquad (1)$$

mit C — Gesamtfläche, q — relativer Flächenanteil Linie 1, Δ_1 — Differenz der Linienpositionen, μ_2 — Position von Linie 2, σ_1, σ_2 — Halbwertsbreiten von Linien 1 bzw. 2. Hierbei gilt Index 1 für die Linie mit niedrigerer, Index 2 für die Linie mit höherer Frequenz.

Auswerteparameter, die vom Experimentator eingegeben werden, sind:

NSPEKT	Spektrennummer
DATUM	Aufnahmedatum
FREQU	Frequenzanfang, Hz
PARAM	Schätzwerte für die Parameter C, μ_2, σ_2, q_1, Δ_1, σ_1
VAR	logische Variable zur Kennzeichnung, ob einer der Signalparameter oder der Nullinienkoeffizienten einen fest vorgegebenen Wert annehmen oder über den Parameterschätzalgorithmus bestimmt werden soll (zehn Werte, davon ein Wert noch für Programmerweiterung verfügbar)
TEXT	Text zur Charakterisierung des Problems

Als Differenz zwischen zwei Spektrenpunkten wird Δ = 0,125 Hz fest vorgegeben. Somit lassen sich die zu den Meßdaten gehörigen Frequenzwerte $X(I)$ im Programm errechnen aus

$$X(I) = \text{FREQU} - 0.125 * (I-1)\ \text{Hz} \qquad (2)$$

$$I = 1,\text{NMESS}$$

Die tatsächliche Zahl der Meßwerte (NMESS, mit NMESS ≦ 2048) wird dadurch ermittelt, daß vom Adapter (vgl. [1]) am Ende der Spektrenaufnahme eine Folge von 15 Nullen erzeugt wird. Im Verarbeitungsprogramm wird der Vektor der Meßwerte auf diese Nullfolge hin abgefragt.

Die Meßwerte $y(x_i)$ und die sich aus (1) ergebenden theoretischen Funktionswerte $f(x_i)$ sind miteinander verknüpft durch

$$y(x_i) = f(x_i) + \varepsilon_i \qquad (3)$$

Für die Meßfehler ε_i wird angenommen, daß sie normalverteilte Zufallsgrößen sind, für die gilt

$$E[\varepsilon(x_i)] = 0 \qquad (4)$$

$$\text{VAR}[\varepsilon(x_i)] \equiv \sigma_i^2 = \sigma^2 \qquad (5)$$

$$\text{COV}[\varepsilon(x_i), \varepsilon(x_j)] = 0 \qquad (6)$$

Die Annahme (6) trifft für NMR-Spektren nicht streng zu. In der Regel findet man jedoch eine rasch abfallende Autokovarianz des Rauschens [6]. Die Abtastgeschwindigkeit von 1 Hz/s wurde so gewählt, daß keine Signalverzerrungen durch Meßwertkorrelation auftreten. Somit kann von der Gültigkeit von (6) ausgegangen werden, was die Anwendung der Methode der (ungewichteten) kleinsten Quadrate zur Parameterschätzung gestattet.

Der hier verwendete Algorithmus wird an anderer Stelle ausführlich dargestellt [7, 8]. Damit der Vorzug des in (1) verwendeten Orthogonalpolynoms zum Tragen kommt, muß programmintern das Frequenzintervall ($x_1 \ldots x_{\text{NMESS}}$), welches nach (2) berechnet wurde, transformiert werden auf (+1 ... −1). Weiterhin wird die Ordinate so skaliert, daß $y_{\max} = 1$ gilt. Dadurch liegen alle Linienparameter im Bereich (−1 ... 2). (Ohne diese Maßnahme hätte C die Größenordnung von 10^3). Die Vorgabe fester Schranken für die gesuchten Parameter wirkt sich auf die Konvergenz des Suchverfahrens günstig aus.

Die Resultate, die der Experimentator erhält, sind Schätzwerte für die Parameter sowie deren Standardabweichungen (rücktransformiert auf die ursprünglichen Koordinatenachsen). Über Drucker werden ausgegeben: ein Protokoll des Ablaufs der iterativen Parameterschätzung, Parameterschätzwerte (einschließlich Koeffizienten für Nullinienpolynom) und deren Standardabweichung vor und nach Rücktransformation, Meßdaten $y(x_i)$ und die Funktionswerte $f(x_i)$, die mit den Parameterschätzwerten berechnet werden. Die graphische Darstellung (DIGIGRAF) zeigt das experimentelle und das theoretische Spektrum sowie die Differenz zwischen beiden, welche zur Beurteilung des Analysenergebnisses wichtig ist.

Je Problem wird das Verarbeitungsprogramm LINFOR nur einmal aufgerufen. Das Schema dieser linearen Problembearbeitung zeigt Abb. 3.

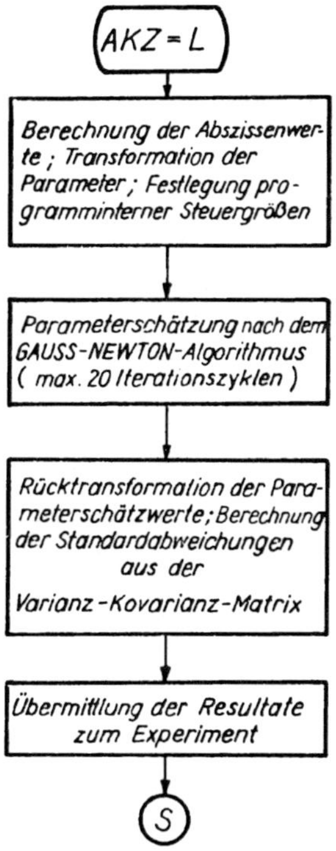

Abb. 3. Programm LINFOR: Ablaufplan

Verarbeitungsprogramm AKKU

Bei der Produktion deuterierter Lösungsmittel ist es nötig, für die Reinheitskontrolle auch kleinste Verunreinigungen nachzuweisen. Zur Erhöhung der Spektrometerempfindlichkeit kann die Eigenschaft einer Spektrenakkumulation dienen, das Signal-Rausch-Verhältnis um den Wert $\sqrt{n_A}$ zu verbessern, wobei n_A die Zahl der akkumulierten Spektren (scans) darstellt. Technische Voraussetzung für die tatsächliche Nutzung der durch die Akkumulation gegebenen Verbesserung der Analysenmöglichkeit ist die Stabilität der Spektrenparameter, insbesondere der Nullinie, über mehrere Stunden [1]. Da die Linienerkennung automatisch, ohne Vorgabe möglicher Lagebereiche erfolgt, ist eine völlig horizontale Nullinie erforderlich, um auch sehr kleine Signale zu finden. Bei dem verwendeten Spektrometer ist diese Bedingung jedoch nicht immer erfüllt, vielmehr weist die Nullinie öfters einen Trend auf, der sowohl ansteigend als auch abfallend sein kann. Die rechnerische Korrektur der Nullinie ist somit von erstrangiger Bedeutung.

Für die Form der Signale wird bei diesem Auswerteprogramm kein analytischer Ausdruck vorgegeben. Zur Erkennung eines Signals bei korrigierter Nullinie dienen die Meßwerte sowie die daraus numerisch ermittelte 1. Ableitung des Spektrums [8]. Um bei der Nullinienkorrektur im Spektrum Signalbereiche von Nullinienbereichen zu unterscheiden, wird lediglich die 1. Ableitung verwendet.

Als Auswerteparameter werden vom Nutzer eingegeben:

NAKK	Zahl der Akkumulationen
NSPEKT	Spektrennummer
DATUM	Aufnahmedatum
FREQU	Frequenzanfang
VSIGN	Frequenz des Vergleichssignals
FREIN	Frequenzabstand zweier Meßdaten

Die Werte FREQU und FREIN (0,5 Hz bei 1000 Hz Meßbereich) dienen zur Skalierung der Abszisse, um Lage und Breite der gefundenen Linien angeben zu können. Der i-te Abszissenwert x_i ergibt sich aus

$$X(I) = \mathrm{FREQU} - \mathrm{FREIN} * (I - 1) \qquad (7)$$

$$I = 1,\mathrm{NMESS}$$

Die Zahl der Meßdaten (NMESS) wird wie bei LINFOR ermittelt.

In [1] wurde ausgeführt, daß die Akkumulation entweder im Mikrorechner oder im Großrechner erfolgen kann. Akkumulation im Mikrorechner wird in AKKU formal wie NAKK = 1 behandelt, der übergebene Meßdatenblock gelangt sofort zur Endauswertung. Für die Akkumulation im EC 1040 ist im COMMON-Bereich zusätzlich Platz für 2048 Werte vom Typ INTEGER * 4 vorgesehen. Außerdem enthält der COMMON-Bereich noch eine Zählgröße K für die Steuerung der Akkumulation, die im aufrufenden Programm zu Beginn der Akkumulatin gleich Null gesetzt wird. Abb. 4 zeigt den Ablauf der Akkumulation.

Bei jedem Aufruf von AKKU wird ein neuer Meßdatenblock übergeben, während die Auswerteparameter unverändert bleiben.

Die Hauptschritte der Auswertung des akkumulierten und gemittelten Spektrums Y sind Glättung, Nullinienkorrektur (mit Bestimmung der Rauschkonstanten), Linienerkennung und -charakterisierung sowie Flächenbestimmung.

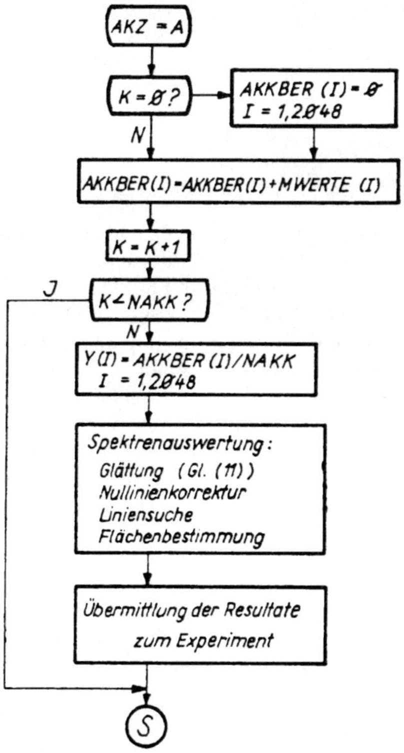

Abb. 4. Programm AKKU: Ablaufplan

MWERTE(I) Meßwerte (Spektrum)
AKKBER(I) Akkumulationsbereich (akkumuliertes Spektrum)
Im aufrufenden Programm NVPNMR wird vor der 1. Akkumulation K = 0 gesetzt

Die Rauschkonstanten R und RS, aus denen die Schwellwerte SW und SWS für die Nullinienkorrektur und die Liniensuche abgeleitet werden, werden aus einem signalfreien Abschnitt am Beginn des Spektrums ermittelt. Um mögliche Artefakte in den ersten Meßdaten zu übergehen, wird der Bereich $I = 16, \ldots, 55$ verwendet. Die Konstante R, die den Rauschpegel charakterisiert, ergibt sich aus

$$R = (y_{\max} - y_{\min}) \qquad (8)$$

während RS (Charakterisierung des Anstiegs im Rauschen) ermittelt wird aus

$$RS = \max |y'| \qquad (9)$$

wobei der Anstieg y_i' berechnet wird aus

$$y_i' = y_{i+1} - y_{i-1} \qquad (10)$$

Vor der Nullinienkorrektur erfolgt eine Glättung des Spektrums. Zur Berechnung eines geglätteten Wertes $\bar{y}_i$ wird eine Ausgleichsparabel durch fünf Meßwerte verwendet [4]:

$$\bar{y}_i = \frac{1}{35}(-3y_{i-2} + 12y_{i-1} + 17y_i + 12y_{i+1} - 3y_{i+2}) \qquad (11)$$

Die Nullinienkorrektur selbst ist ein iterativer Prozeß. Auf der Basis eines Anstiegsschwellwertes SWS, der aus RS durch

$$SWS = RS * 1.25 \qquad (12)$$

gebildet wird (wobei der Faktor 1,25 empirisch festgelegt worden ist), werden diejenigen Spektrenbereiche herausgesucht, welche als Linien anzusehen sind. Die verbleibenden Abschnitte werden als linienfrei betrachtet; um zu vermeiden, daß diese Abschnitte auch Bereiche links und rechts von einer Linie enthalten, die durch Spektrometereffekte beeinflußt werden (z.B. durch Rotationsseitenbänder), werden die gefundenen Linienbereiche links vom jeweiligen Anfang LA und rechts vom Ende LE verbreitert um

$$D = (LE - LA) * (2/3) \qquad (13)$$

Durch die linienfreien Abschnitte wird nach der Methode der kleinsten Quadrate ein Polynom 4. Grades gelegt, welches vom Spektrum subtrahiert wird. Der Zyklus wird mehrfach durchlaufen, da sich durch die Korrektur das Ausmaß der Nullinienbereiche ändert; die Erfahrung mit Testbeispielen hat gezeigt, daß nach fünf Iterationen Konvergenz eintritt.

Das geglättete und korrigierte Spektrum wird anschließend einer Liniensuche unterworfen. Da eine horizontale Nullinie vorausgesetzt wird, kann nicht nur ein Anstiegsschwellwert, sondern auch ein Schwellwert SW für die Spektrenamplitude verwendet werden. Zu Beginn der Liniensuche wird (empirisch) angesetzt

$$\left.\begin{aligned} SW &= R * 0.75 \\ SWS &= RS * 0.75 \end{aligned}\right\} \qquad (14)$$

Übersteigt bei der Abarbeitung des Spektrums die Zahl der gefundenen Linien den Wert 40, so wird die Liniensuche mit vergrößerten Schwellwerten

$$\left.\begin{aligned} SW &= SW * 1.5 \\ SWS &= SWS * 1.5 \end{aligned}\right\} \qquad (15)$$

wiederholt, da angenommen werden muß, daß zu viele Zufallslinien gefunden worden sind. Für jede Linie J wird gespeichert

$LA(J)$	Linienanfang
$LE(J)$	Linienende
$LP(J)$	Linienposition (Abszissenwert mit maximaler Spektrenamplitude)
$LSCH(J)$	Position einer Schulter

Der verwendete Algorithmus trennt die Linien einer Liniengruppe durch Lotfällung an der Stelle des Amplitudenminimums bzw. in der Mitte eines Bereichs mit $y' < SWS$ und $y > SW$.

Bei der abschließenden Flächenbestimmung wird über das Spektrum jeweils von $LA(J)$ bis $LE(J)$ integriert. Verwendet wird die Methode der Berechnung bestimmter Integrale nach der Trapezformel für gegebene Funktionswerte bei gleichen Abszissendifferenzen Δ (= FREIN):

$$F(J) = \int_{x_1}^{x_i} y(x)\,\mathrm{d}x \approx \sum_{K=IA(J)+1}^{IE(J)} \frac{\Delta}{2}(y_k + y_{k-1}) \qquad (16)$$

Weiterhin werden die Flächen auf die des Vergleichssignals bezogen. Als Vergleichssignal wird diejenige Linie verwendet, deren Position LP der vorgegebenen Frequenz VSIGN am nächsten liegt.

Als Resultate werden an den Ort des Experiments zurückgegeben: $LA(J)$, $LE(J)$, $LP(J)$ sowie die Flächen, die relativen Flächen und die

Amplitude an der Stelle $LP(J)$. Das Druckerprotokoll enthält neben allen Linienparametern auch das (unkorrigierte) akkumulierte Spektrum. Schließlich ist die graphische Darstellung des akkumulierten Spektrums vor und nach der Nullinienkorrektur vorgesehen.

Verarbeitungsprogramm ISOTAN

In der Isotopenanalyse nimmt die NMR-Spektroskopie einen besonderen Platz ein, weil mit dieser Meßmethode sowohl die Position der Isotopenmarkierung als auch die Höhe der Anreicherung am unzerstörten Molekül bestimmt werden kann [9]. Das verwendete Spektrometer ist für eine Isotopenanalyse am Wasserstoff geeignet.

Über den Einsatz der NMR zur Bestimmung des Deuteriumgehaltes in gelösten organischen Verbindungen liegen mehrere zusammenfassende Berichte vor [9, 10]. Bewährt hat sich die Verwendung eines inneren Standards, der ebenso wie die zu untersuchende Substanz in das Proberöhrchen eingewogen wird. Die erforderliche Substanzmenge beträgt 0,05 ... 0,5 ml. Zur Vermeidung von systematischen Fehlern ist der Einsatz von zwei inneren Intensitätsstandards vorteilhaft[2]).

Die Fläche unter dem Resonanzsignal ist proportional der Zahl der Atome des jeweiligen Isotops in der betrachteten chemischen Gruppe und hängt unter geeigneten apparativen Bedingungen nicht von Strukturparametern der Substanz ab. Die Fläche der Resonanzsignale der zu untersuchenden Substanz sowie der Standards kann für eine Handauswertung mittels des elektronischen Integrators am Spektrometer direkt gemessen werden. Bei der hier dargelegten automatisierten Auswertung werden Linienpositionen und Flächen vom Originalspektrum ausgehend bestimmt.

Der Deuteriumgehalt D_i der chemischen Gruppe i bezogen auf die Referenzsubstanz r kann aus folgender Beziehung ermittelt werden:

$$D_{ir}\,[\%] = 100 - \frac{M^{D}_{Sub_r} \cdot m_{Ref_r} \cdot N_{Ref_r} \cdot F_i}{m_{Sub} \cdot M_{Ref_r} \cdot N_i \cdot F_{Ref_r}} \cdot 100 \quad (17)$$

Hierbei sind

m_{Sub} – Masse der untersuchen Probensubstanz
m_{Ref_r} – Masse der verwendeten Referenzsubstanz r
M_{Ref_r} – Molmasse der Referenzsubstanz r
N_i – Zahl der Protonen in Gruppe i (i = 1,NGR)
N_{Ref_r} – Zahl der Protonen in Referenzsubstanz r
F_i – Signalfläche der Gruppe i
F_{Ref_r} – Signalfläche der Referenzsubstanz r

Die Molmasse M^{D}_{Sub} der teildeuterierten Substanz, berechnet aus der Referenzsubstanz r, ergibt sich aus

$$M^{D}_{Sub_r} = \frac{M_{Sub} + 1{,}0063 \cdot \sum_{i=1}^{NGR} N_i}{1 + \dfrac{1{,}0063 \cdot m_{Ref_r} \cdot N_{Ref_r} \sum_{i=1}^{NGR} F_i}{M_{Ref_r} \cdot m_{Sub} \cdot F_{Ref_r}}} \quad (18)$$

mit

M_{Sub} – Molmasse der undeuterierten Substanz.

[2]) Die Isotopenanalyse kann auch ohne Verwendung eines inneren Standards durchgeführt werden, indem die deuterierte Substanz in einem NMR-Röhrchen spektroskopiert wird, an dessen Boden eine geschlossene Kapillare von ca. 6 cm Länge konzentrisch angeschmolzen ist, welche eine Referenzsubstanz enthält. Der Deuteriumgehalt der Gruppe i läßt sich dann aus der Formel $(100 - (A \cdot M_{Sub} \cdot F_i \cdot 100)/(\varrho_{Sub} \cdot N_i \cdot F_{Ref}))$ ermitteln, wobei A ein für das betreffende Proberöhrchen zu ermittelnder Eichfaktor und ϱ_{Sub} bzw. M_{Sub} die Dichte bzw. Molmasse der unmarkierten Substanz darstellen.

Falls die markierte Verbindung in schwerem Wasser gelöst ist und eine Gruppe i mit schnell austauschenden Wasserstoffatomen enthält, so erhält man für die Gruppe i kein getrenntes Signal. Man beoachtet nur die Summe F_{S_i} des Signals der Gruppe i und des Restsignals vom Wasser. Die für die Berechnung des Deuterierungsgrades benötigte Signalfläche F_i kann ermittelt werden über die Beziehung

$$F_i = \frac{\dfrac{M_{D_2O} \cdot N_i}{M^{D}_{Sub} \cdot N_{D_2O}}}{\dfrac{1 - c/100}{c/100} + \dfrac{M_{D_2O} \cdot N_j}{M^{D}_{Sub} \cdot N_{D_2O}}} \cdot F_{S_i} \quad (19)$$

Mit c wird die Gewichtskonzentration der gelösten Substanz in Gew.-% bezeichnet. Eine Besonderheit bei dieser Variante ist, daß grundsätzlich nur eine Referenzsubstanz verwendet wird. Der Deuterierungsgrad der Gruppe i folgt somit aus

$$D_i\,[\%] = 100 - \frac{100 \cdot m_{Ref} \cdot N_{Ref} \cdot F_{S_i} \cdot 100}{c \cdot m_{Sub} \cdot M_{Ref} \cdot N_i \cdot F_{Ref}} \times \frac{M_{D_2O} \dfrac{N_i}{N_{D_2O}}}{\dfrac{100 - c}{c} + \dfrac{M_{D_2O} \cdot N_i}{M^{D}_{Sub} \cdot N_{D_2O}}} \quad (20)$$

Für die Molmasse M^{D}_{Sub} kann vereinfacht

$$M^{D}_{Sub} = M_{Sub} + 1{,}0063 \cdot \sum_{i=1}^{NGR} N_i \quad (21)$$

eingesetzt werden.

Werden Messung und Auswertung mehrfach durchgeführt (k = 1,NME), so lassen sich Mittelwert und statistischer Fehler der Flächen- und D-Gehalt-Bestimmungen aus den allgemeinen Formeln

$$\bar{x} = \sum_{k=1}^{NME} x_k/\mathrm{NME} \quad (22)$$

und

$$S_x = \frac{\sqrt{\sum_{k=1}^{NME} (x_k - \bar{x})^2}}{\mathrm{NME(NME} - 1)} \cdot t_{NME} \quad (23)$$

ermitteln, wobei t_{NME} den Wert der t-Verteilung für die Irrtumswahrscheinlichkeit $\alpha = 0{,}05$ darstellt.

Zum Vergleich lassen sich die Fehler der Flächenbestimmung auch über eine Eichkurve angeben. Diese Eichkurve resultiert aus den bisherigen Erfahrungen beim Betrieb des Spektrometers und gibt einen empirischen Zusammenhang zwischen dem Fehler der Flächenbestimmung und der Protonenkonzentration P_i der entsprechenden Gruppe i. Ausgangspunkt für die Bestimmung dieser Protonenkonzentration ist die Berechnung der Protonenkonzentration der Referenzsubstanz 1 mittels

$$P_{Ref_1} = N_{Ref_1} \cdot \frac{m_{Ref_1}}{M_{Ref_1}} \cdot 2000 \quad \mathrm{mol}/l \quad (24)$$

da das Probenvolumen ca. 0,5 ml beträgt. Die übrigen Protonenkonzentrationen ergeben sich aus P_{Ref_1} durch Multiplikation mit den Flächenverhältnissen

$$P_i = P_{Ref_1} \cdot \frac{F_i}{F_{Ref_1}} \quad (25)$$

und die Gesamtprotonenkonzentration der Substanz aus

$$P = P_{Ref_1} \cdot \frac{\sum_{i=1}^{NGR} F_i}{F_{Ref_1}} \quad (26)$$

Die Abhängigkeit des Fehlers der Fläche i von der Protonenkonzentration P_i wird empirisch dargestellt durch

$$\varDelta_i = \frac{23}{\log\,(1000\ P_i)\cdot\sqrt{\mathrm{NME}}} \quad [\%] \tag{27}$$

Für den Fehler der D-Gehalt-Bestimmung ergibt sich ebenfalls ein Vergleich zum statistischen Fehler aus (23) durch

$$\lambda_i = (100 - D_i)\cdot 0{,}1 \quad [\text{at-}\%] \tag{28}$$

Diese beiden Vergleiche lassen erkennen, ob beim Meß- und Auswerteprozeß systematische Fehler aufgetreten sind.

Auswerteparameter für die Nutzung des Programms ISOTAN sind

NSPEKT	
DATUM	
FREQU	
NSUB	Substanznummer
$M(I)$, $I = 1{,}3$	Substanzeinwaagen
VAR(I), $I = 1{,}2$	logische Variable

Die Abszisse wird gemäß Gl. (7) skaliert, wobei nur der 1000-Hz-Bereich zulässig ist. Der Vektor $\boldsymbol{M}$ enthält die Massen der Substanzprobe und der beiden Referenzsubstanzen. Die erste der beiden logischen Variablen legt fest, ob bei der Berechnung des D-Gehalts Variante 1 (Gl. (17)) oder Variante 2 (Gl. (200)) gewählt wird. Bei Variante 2 (nur eine Referenzsubstanz) enthält $M(3)$ die Gewichtskonzentration c der gelösten Substanz in Gew.%. Die zweite logische Variable bestimmt den Umfang des auf dem Drucker ausgegebenen Protokolls.

Die Substanznummer NSUB ermöglicht die Nutzung einer permanenten Direktzugriffsdatei. Jeder Satz dieser Datei enthält Angaben zu einer bestimmten Kombination von Testsubstanz und Referenzsubstanzen:

MSUB	Molmasse der undeuterierten Substanz
NGR	Zahl der Gruppen
NP(I), $I = 1$,NGR	Zahl der Protonen je Gruppe
NRE	Zahl der Referenzsubstanzen
MREF(R), $R = 1$,NRE	Molmassen der Referenzsubstanzen
NREF(R), $R = 1$,NRE	Zahl der Protonen je Referenzsubstanz
FRU(I), FRO(I)), $I = 1$,NGR	Position des Signals der Gruppe i (untere und obere Grenze des in Betracht kommenden Frequenzintervalls)
(FRUREF(R), FROREF(R)), $R = 1$,NRE	Position des Signals der Referenzsubstanz r
TEXT	Name der Substanz und der Referenzen

Bei der Verarbeitung des ersten Spektrums einer Meßserie ($k = 1$) wird der durch NSUB gekennzeichnete Satz gelesen, und die Daten stehen dann für die folgenden Berechnungen zur Verfügung.

Der Programmablaufplan ist in Abb. 5 dargestellt. Es sind drei verschiedene Verarbeitungskennzeichen möglich: I, Y oder X. Vor dem ersten Programmaufruf wird ein Programm NVPNMR der Spektrenzähler $K = 0$ gesetzt (Variable im COMMON-Bereich). Verarbeitungskennzeichnen I bedeutet, daß das gesendete Spektrum ausgewertet werden soll. (Im Gegensatz zum Problem AKKU wird ein Spektrum nach Aufnahme nicht automatisch zum Großrechner gesendet, sondern der Experimentator hat die Möglichkeit, eine Messung unmittelbar nach Spektrenaufnahme zu verwerfen.) Spektrenauswertung bedeutet Bestimmung der relevanten Signalflächen und die Berechnung des D-Gehalts gemäß Gl. (17) (VAR(1) = T) oder Gl. (20) (VAR(1) = F). Anschließend werden die aus dem einzelnen Spektrum ermittelten Resultate an den Experimentator zurückgegeben und ermöglichen so die Festlegung des nächsten Schrittes. Entweder erfolgt mit AKZ = I eine neue Spektrenaufnahme mit anschließender Auswertung, oder das letzte Resultat wird mit AKZ = X verworfen: Zurückstellen des Spektrenzählers und Löschen der Zwischenresultate. — Ist eine hinreichende Anzahl von Spektren aufgenommen (max. 8), so kann mit AKZ = Y die Endauswertung aufgerufen werden: Bildung von Mittelwerten für Flächen und D-Gehalte (einzelne Gruppen und Gesamt-D-Gehalt) sowie Berechnung der Fehler.

Liniensuche und Flächenbestimmung erfolgen ausgehend vom Originalspektrum und geschehen in der gleichen Weise wie im Verarbeitungsprogramm AKKU. Resultate dieses Arbeitsschrittes sind eine Folge von Linien, gekennzeichnet durch die Linienposition $LP(J)$ und die Fläche $F(J)$. Einer bestimmten Gruppe i werden alle Linien j (und damit deren Flächen) zugeordnet, für die

$$\mathrm{FRU}(I) \leq LP(J) \leq \mathrm{FRO}(I) \tag{30}$$

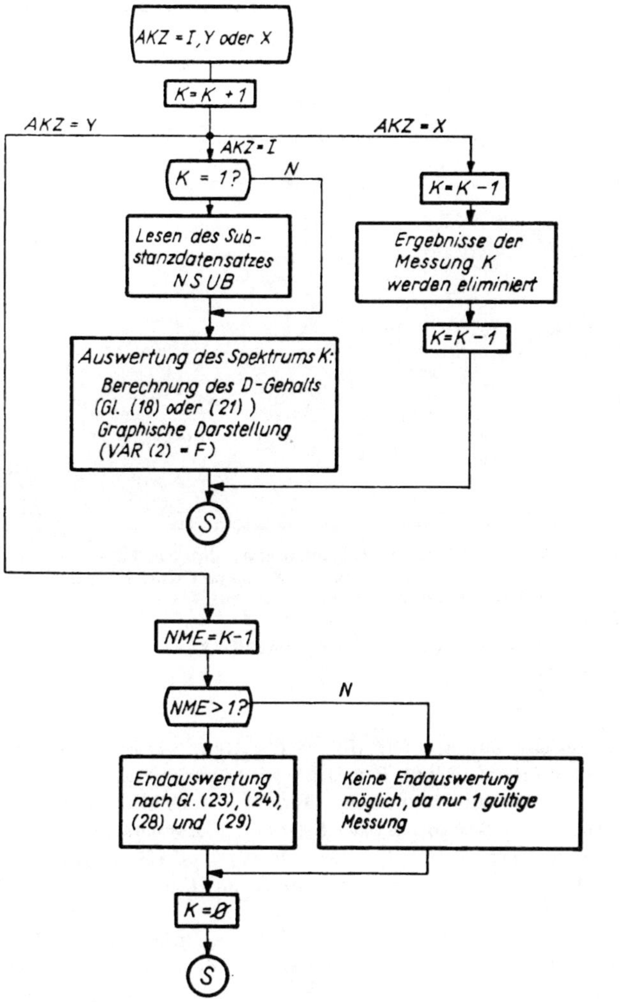

Abb. 5. Programm ISOTAN: Ablaufplan
Im aufrufenden Programm NVPNMR wird vor dem 1. Aufruf K = 0 gesetzt

gilt ($I = 1$,NGR; $J = 1$,NL). In ähnlicher Weise wird bei den Referenzsubstanzen verfahren.

Zwischen- und Endergebnisse werden an den Ort des Experiments zurückgesendet. Über Drucker wird ein Protokoll der Auswertung ausgegeben, dessen Umfang durch VAR(2) bestimmt wird. VAR(2) = F bedeutet auch eine graphische Darstellung der einzelnen Spektren (jeweils vor und nach der Nullinienkorrektur, zusätzlich das Spektrenintegral).

Beispiele

Abb. 6 zeigt ein Resultat, das mit dem Programm LINFOR gewonnen wurde. Die Linienformanalyse wurde am $N(CH_3)_3^+$-Signal der Kopfgruppen von Doppelschichtliposomen in D_2O vorgenommen. Das Spektrum wurde 96 Tage nach Zugabe von 15 mM $EuCl_3$ mit einer Abtastgeschwindigkeit von 1 Hz/s aufgenommen (Details in [2, 11]). Eingangsparameter waren

$$C = 10\,000 \qquad \mu_2 = 299\ \mathrm{Hz} \qquad \sigma_2 = 4{,}0\ \mathrm{Hz}$$
$$q_1 = 0{,}66 \qquad \varDelta_1 = 4\ \mathrm{Hz} \qquad \sigma_1 = 4{,}0\ \mathrm{Hz}$$

Bereits nach fünf Iterationen waren die Endwerte erreicht:

$$C = 5662 \pm 58 \qquad \mu_2 = 299{,}1 \pm 1{,}7$$
$$\sigma_2 = 1{,}87 \pm 0{,}07 \qquad q_1 = 0{,}751 \pm 0{,}008$$
$$\varDelta_1 = 5{,}56 \pm 0{,}04 \qquad \sigma_1 = 2{,}66 \pm 0{,}04$$

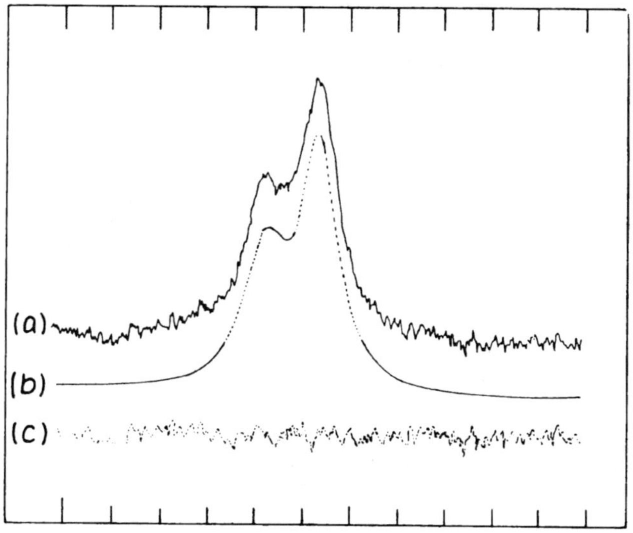

Abb. 6. Testbeispiel LINFOR: Auswertung des $N(CH_3)_3^+$-Signals der Kopfgruppen von Doppelschichtliposomen in D_2O bei Zugabe von 15 mM $EuCl_3$

a) experimentelles Spektrum
b) theoetisches Spektrum (ca. 1 cm verschoben)
c) Differenzspektrum aus a) und b)

Die Reststreuung σ (Gl. (5)) betrug 13,0, was auf eine gute Anpassung hinweist. Bestätigt wird diese Aussage durch die graphische Darstellung. Kurve (a) zeigt das experimentelle Spektrum, Kurve (b) das theoretische Spektrum (um ca. 1 cm nach unten verschoben), und Kurve (c) stellt die Differenz zwischen Experiment und theoretischer Analyse dar.

Das Demonstrationsbeispiel für das Programm AKKU (und damit für die Effizienz des Liniensuchalgorithmus, wie er auch beim Programm ISOTAN zum Einsatz kommt) diente dem Nachweis einer geringen Verunreinigung bei 4,4 ppm in Nitromethan-d_3 (99,7 at-% D) durch Spektrenakkumulation. Das in Betracht kommende Teilspektrum

Tab. 1. Testbeispiel Akkumulation: Resultate der Liniensuche mit einer Akkumulation

Linienparameter, Hz			Position, ppm	relative Fläche	Amplitude
Anfang	Position	Ende			
192,0	187,5	181,0	2,33	1,0	5078
179,0	171,0	163,0	2,14	1,83	5060

Frequenzanfang: 390 Hz, Frequenzende: 145 Hz
Frequenz Vergleichssignal: 189 Hz

Tab. 2. Testbeispiel Akkumulation: Resultate der Liniensuche nach 250 Akkumulationen

Linienparameter, Hz			Position, ppm	relative Fläche	Amplitude
Anfang	Position	Ende			
356,5	353,5	346,0	4,42	0,03	127
212,5	208,5	205,0	2,61	0,04	269
198,0	188,5	185,0	2,35	1,0	5588
185,0	184,5	182,5	2,31	0,04	686
182,5	173,0	164,0	2,16	1,71	4878
164,0	163,5	161,5	2,04	0,005	170

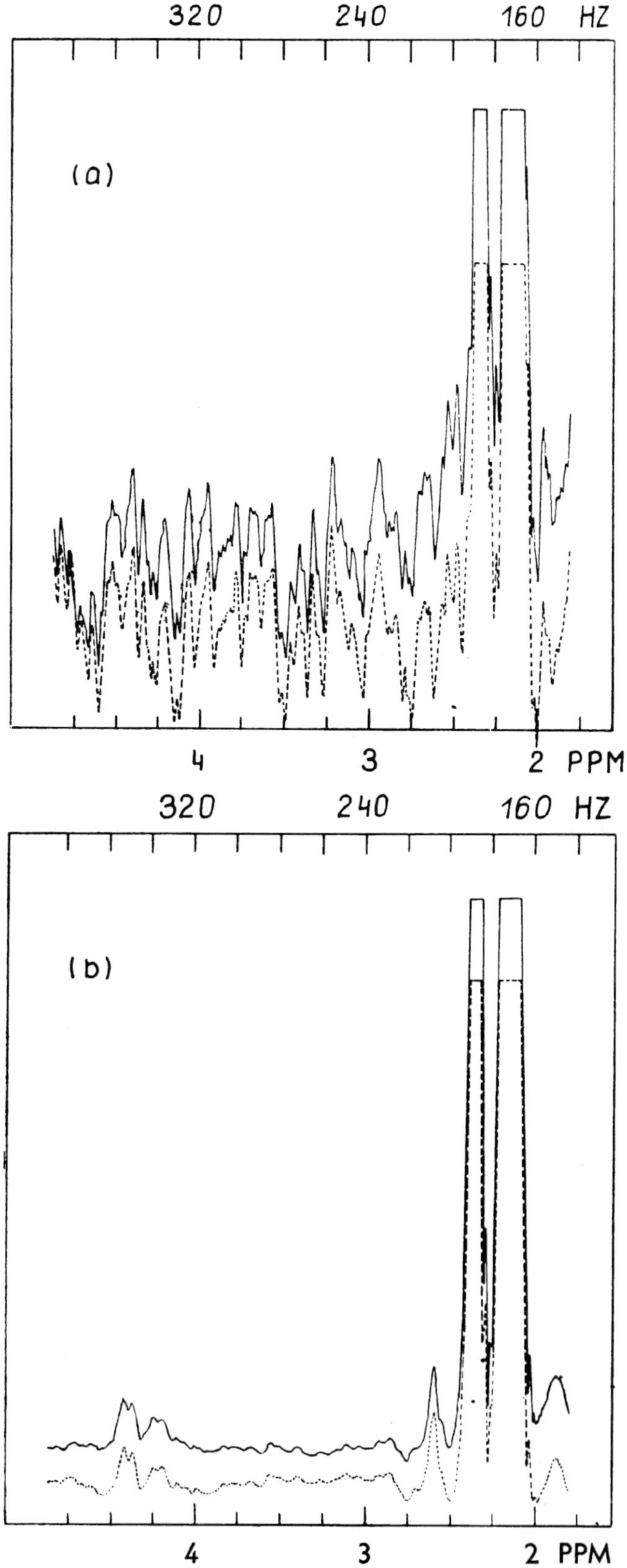

Abb. 7. Testbeispiel AKKU: Nachweis einer geringen Verunreinigung bei 4,4 ppm in Nitromethan-d_3 (99,7 at.-%) durch Spektrenakkumulation

a) 1 scan
b) 250 scans

Die obere Kurve stellt jeweils das experimentelle Spektrum, die untere das Spektrum nach Nullinienkorrektur dar wurde mit einer Abtastgeschwindigkeit von 20 Hz/s über das mikrorechnerintensive Akkumulationsprogramm [1] aufgenommen. Abb. 7 zeigt das Spektrum ohne Akkumulation (1 scan, Tab. 1, Abb. 7a) und nach 250 scans (Abb. 7b), jeweils ohne (oberes Spektrum) und mit Nulllinienkorrektur (unteres Spektrum).

Im Frequenzintervall 390 ... 144 Hz wurden mit 1 scan zwei Linien gefunden, das Substanz- und das Vergleichssignal. Signale von Verunreinigungen gingen im Rauschen unter. Nach 250 Akkumulationen hob sich das Signal der vermuteten Verunreinigungen sichtbar aus dem Rauschen heraus (Tab. 2, Abb. 7b). Zur Beurteilung des Resultats sei angegeben, daß die gleiche Substanz an einem WP-80-Spektrometer (BRUKER) nach 600 scans für den Peak bei 4,4 ppm ein Signal-Rausch-Verhältnis von 5:1 lieferte.

Literatur

[1] *B. Lindemann, H. Sprinz, U. Hädrich, R. Jung, G. Sack,* Isotopenpraxis 22 (1986) 32

[2] *H. Sprinz, E. Winkler, H. Schäfer,* Isotopenpraxis 17 (1981) 171

[3] *E. Winkler,* unveröffentlichte Resultate

[4] *R. Zurmühl,* Praktische Mathematik für Ingenieure und Physiker. Berlin: Springer-Verlag 1965

[5] *G. A. Pearson,* J. Magn. Res. 27 (1977) 265

[6] *D. Ziessow,* On-line-Rechner in der Chemie. Berlin: Walter de Gruyter 1973

[7] *E. Winkler,* Biom Z. 16 (1974) 259

[8] *E. Winkler,* Dissertation B. Leipzig 1982

[9] *G. Müller, K. Mauersberger,* Analyse stabiler Isotope durch spezielle Methoden, Berlin: Akademie-Verlag 1969

[10] *G. J. Martin, M.-I. Quemeneur, M. L. Martin,* Org. Magn. Res. 1 (1969) 191

[11] *Z. V. Kuropteva, H. Sprinz, H. Schäfer, E. Winkler,* ZfI-Mitt. 102 (1985) 21

Isotopenpraxis 22 (1986) 11, S. 407 bis 410

Verification of the Usefulness of Radiotracing Methods in Hydrodynamic Investigations

A. G. Chmielewski, A. Dobrowolski

(Institute of Nuclear Chemistry and Technology, Warszawa, Poland)[1]

In the paper comparison of accuracy of two methods used for investigations of hydrodynamic in continuous flow tanks has been made. Basis for the comparison is the experimental work performed for big industrial wastewater concentration equalization tank. The first method is based on the radiotracer experiment while the second one uses the analysis of response on sudden change of the technological parameters. The radiotracer method has better accuracy connected with the shorter sampling period.

Es wird die Genauigkeit zweier Methoden zur Untersuchung der Hydrodynamik von kontinuierlich arbeitenden Flüssigkeitsbehältern verglichen. Grundlage dafür sind experimentelle Untersuchungen an großen Industrieabwasser-Ausgleichsbehältern. Die erste Methode bedient sich der Radiotracertechnik, während die zweite Methode auf einer plötzlichen Veränderung der technologischen Parameter beruht. Es zeigt sich, daß die Radiotracermethoden die besseren Ergebnisse liefert, wobei außerdem noch kürzere Probenahmeperioden von Vorteil sind.

Keywords

bromine 82; flow models; frequency response; hydrodynamical model; hydrodynamics; tracer techniques; water pollution

1. Introduction

Sometimes from the side of specialists in the field of process engineering some objections are being rised concerning the trustworthiness of radiotracing methods used in process investigations. While rising such objections the influence of such effects as diffusion resulting in axial and radial dispersion, adsorption by apparatus wall and suspended matter etc. is frequently being pointed out. These effects are said to be responsible for worse accuracy of radiotracing investigation results. Nevertheless the practice shows that such objections are not true ones. Many practicians, however, using the radiotracing methods are accepting them "a priori" as perfect ones. Very rarely the results of trade-off studies on this matter based on the analysis of process variables are being presented. This results mainly from the fact that it is difficult to carry out measurements of changes in process variables within a necessary range being required to make necessary calculations. It is also not always feasible to introduce into process under investigation a disturbance which gives a signal being easier to be analysed.

This paper presents the results of a study which render possible to correlate a model of flow dynamics being derived from radiotracer measurements with that derived from direct measurements of a process variable which immediately characterizes a subject of the study.

As a subject of this study an equalization tank used in a biological treatment plant was taken, whereas the chemical oxygen demand (COD) was a process variable the changes in which were investigated. This variable is a basic quantity being indicative of sewage treatment plant performance.

2. Equalization Tank Under Investigation

A diagramatic drawing of the apparatus being investigated is shown in Fig. 1. It includes an equalization tank of 4950 m^3 cubic capacity. Flow rate of wastes flowing there

[1]) Address: PL-03-195 Warszawa, Dorodna 16, Poland

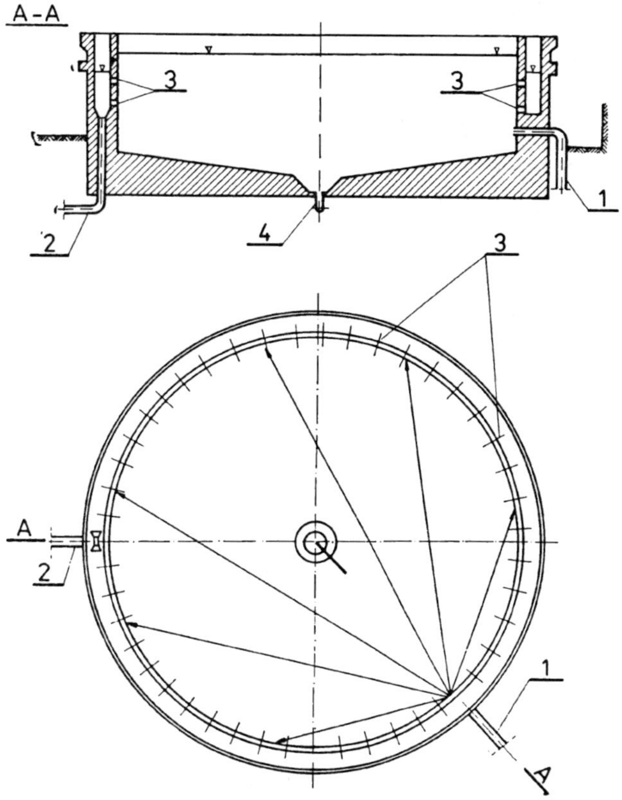

Fig. 1. Diagramatic drawing of the apparatus under investigation

through was of 700 m³/h throughout all the tests being performed.

The wastes were admitted into an internal tank *2* through a pipeline *1* and discharged therefrom through openings disposed at two different levels into an overflow trough *3* and further on through a pipeline *4* into a biological treatment basin. Small amount of sludge being sedimented during operation was periodically discharged from a funnel disposed in the bottom of the tank through a discharge pipeline *5*. The equalization tank was provided with a sludge scraper being not shown in the attached drawing. A thorough analysis of equalization tank operation in conjunction with operation of a complete treatment plant has been presented in another paper being published in a professional journal [1].

3. A Model of Flow Dynamics Derived from Results of Radiotracer Examinations

As radiotracer the Br-82 gamma radioisotope was used. It was injected in the form of aqueous solution of KBr together with a non-active carrier. Since the duration of radiotracer injection was incommensurably short as compared with a time constant of the apparatus under investigation an inlet signal can be considered as having the form of δ-type *Dirac* pulse

$$\delta(t) = \begin{cases} \infty & t = 0 \\ 0 & t \neq 0 \end{cases} \tag{1}$$

Measurements of the inlet signal were carried out using a scintillation probe connected to a scaler and a printer. Data acquisition intervals of 1 min duration were used. The response curve $R(t)$ is shown in Fig. 2.

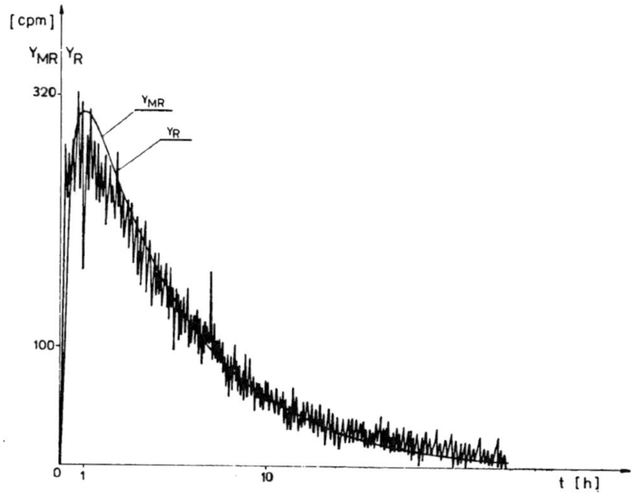

Fig. 2. Response curve for a pulse radiotracer injection

Average residence time $\bar{\tau}_p$ (resulting from process data)

$$\bar{\tau}_p = \frac{V}{q} \quad [\mathrm{h}] \tag{2}$$

where V = cubic capacity of the tank [m³]
q = volumetric flow rate [m³/h]

was about 7 h for the apparatus under investigation.

Since the response curve being obtained at the outlet is a function of residence time distribution then the actual residence time $\bar{\tau}_{rz}$ can be calculated therefrom

$$\bar{\tau}_{rz} = \frac{\int\limits_0^\infty t \cdot R(t)\, dt}{\int\limits_0^\infty R(t)\, dt} \tag{3}$$

where $R(t)$ is a counting rate recorded at the scaler connected to the scintillation probe which was placed at the tank outlet.

In accordance with the *Wolf-Resnick* notation [2] the transmittance of investigated apparatus, and consequently its dynamic model, can be expressed in the following form:

$$G(s) = \frac{\exp(-\bar{\tau}_p \cdot s)}{(1 + \eta_i \cdot \bar{\tau}_p \cdot s) \ldots (1 + \eta_n \cdot \bar{\tau}_p \cdot s)} \tag{4}$$

Using the numerical methods the response curve has been matched with that being obtained for a model described by the general relationship [4]. The best matching of curves has been obtained for a model made up of two arranged in tandem mixing tanks V_1 and V_2 having different time constants T_1 and T_2 with an in-series connected volume V_{pt} where piston flow takes place and a dead space V_m where flow stagnation conditions prevail. This model is shown in Fig. 3.

The transmittance for such a model takes the following form:

$$G(S) = \frac{\exp(-T_t \cdot s)}{(1 + T_1 \cdot s) \cdot (1 + T_2 \cdot s)} \tag{5}$$

The dead space has been derived from a difference between the calculated mean residence time $\bar{\tau}_p$ and the actual mean residence time $\bar{\tau}_{rz}$. The fraction f_{ds} of the total tank volume V, representing this portion thereof where flow stagnation conditions will prevail amounts:

$$f_{ds} = \frac{V_m}{V} = \frac{q(\bar{\tau}_p - \bar{\tau}_{rz})}{q \cdot \bar{\tau}_p} = 1 - \frac{\bar{\tau}_{rz}}{\bar{\tau}_p} \tag{6}$$

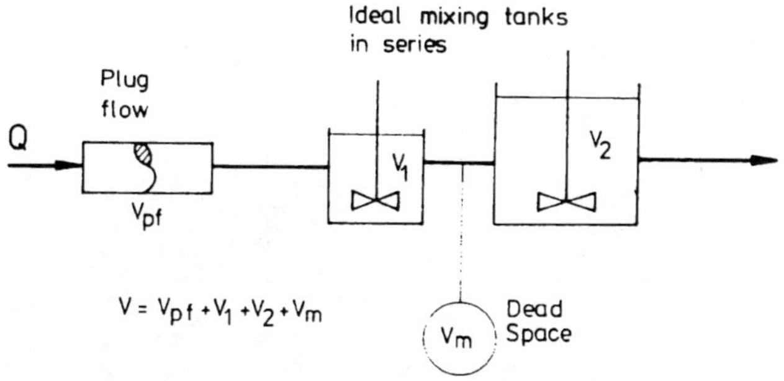

Fig. 3. Model of flow hydrodynamics in the apparatus under investigation

Tab. 1. The model parameters obtained from radiotracer method (A) and from the analyse of COD changes (B). For the symbols look at Fig. 3.

	V_{pf} [m³]	V_1 [m³]	V_2 [m³]	V_m [m³]
A	70	280	3570	990
B	0	280	4060	610

Values of the model variables as obtained from radiotracer examinations have been tabularized in the Tab. 1. The response curve obtained for the model has been plotted, for comparison, together with that being obtained as result of the radiotracing experiments (Fig. 2).

4. A Model of Flow Dynamics Derived from Measurements of Changes in Values of Process Variable

In the course of the radiotracing experiment concurrent measurements of changes in chemical oxygen demand at inlet and outlet of the equalization tank were carried out. During 31 h a 500 ml large sample of waste water was taken every 1/2 h both from inlet and outlet streams. After filtering each sample it was used for determination of COD by the chromate titration method [3]. COD variations at both the inlet and outlet of the equalization tank are shown in Fig. 4. During 11-th hour from the start of measuring sequence an emergency waste dump with high COD load took place. Maximum COD value was of about 36 kg O_2/m^3 at average COD value amounting to about 1·12 kg O_2/m^3. This emergency waste dump was of about 1 h duration and can be considered as a pulse proportioning (of predetermined pulse duration). On the COD variation diagram corresponding portion of the response curve for the equalization tank outlet represents the response to the disturbance caused by this emergency waste dump. This portion of inlet/outlet COD curves is shown separately, in enlarged scale, in Fig. 5.

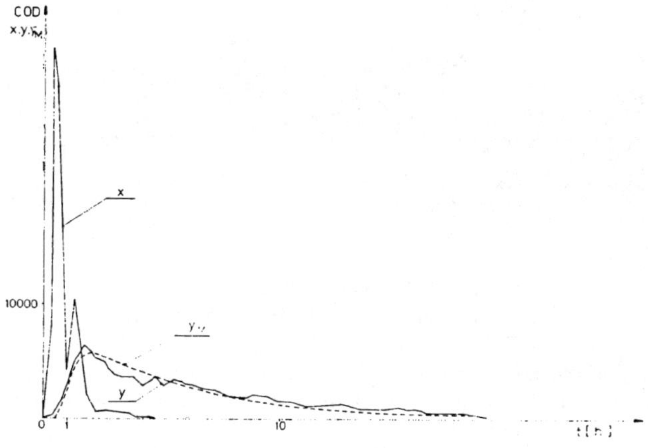

Fig. 5. Portion of COD variation curves subject to analysis

Since in this case the inlet signal can not be considered as an ideal δ-pulse the calculations being necessary for deriving a dynamic model were made in another way. The method of hybrid modelling was used while the matching of model was based on an input convolution with modelled pulse response. The best matching between the radiotracing output curve and the model curve was ob-

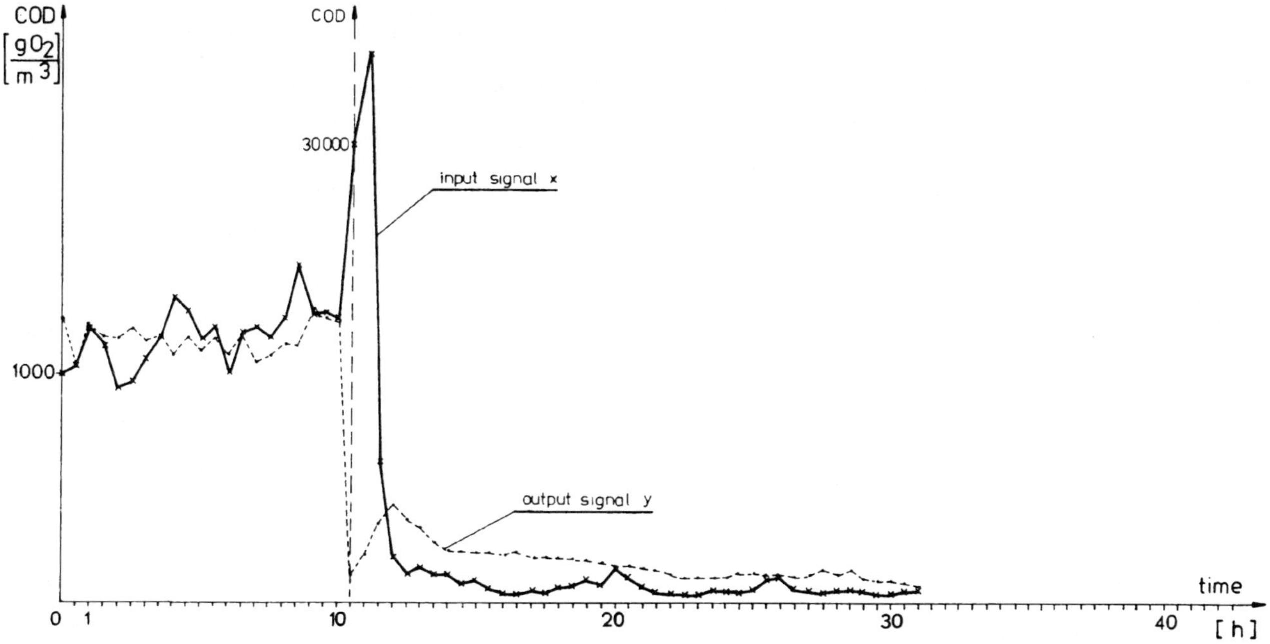

Fig. 4. Variation of COD at inlet and outlet of the apparatus under investigation

tained for a cascade of two mixing tanks having their time constants as shown in the Tab. 1. The model curve is shown in Fig. 5 by dashed line. In order to obtain better visualization of experimental compliance between these two curves (i.e. output and model ones) the output curve has been shown in Fig. 6 (with correlation coefficient of 0.98).

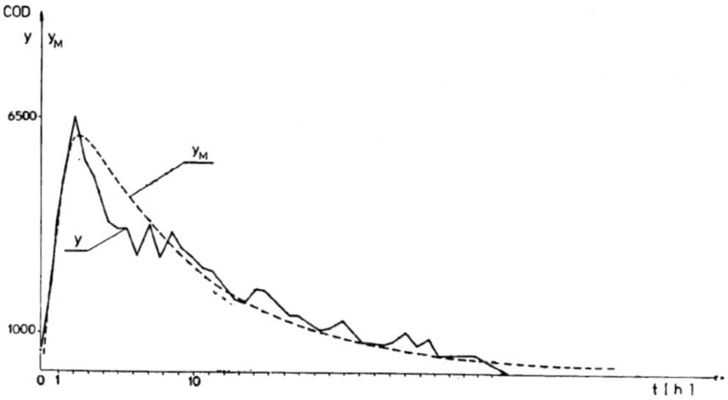

Fig. 6. Comparison of model function of COD variations in an inlet stream with actual function thereof.

5. Comparison of Models Derived by Radiotracing and COD Measurements

A comparison between pulse responses obtained for the model based on radiotracing (curve A) and for that based on determined COD variations (curve B) is made in Fig. 7. Fig. 8 shows a comparison between frequency characteristics for the both models.

The consistence of two curves is very good. Similary there is a very little difference (not exceeding 10%) between residence times obtained for two models. In both cases

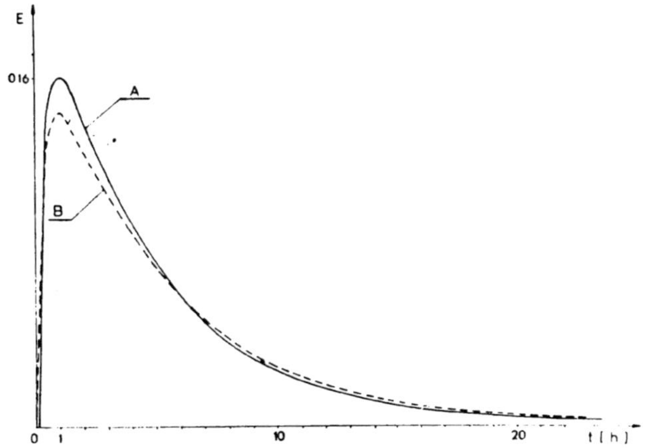

Fig. 7. Comparison of response curves for the radiotracing experiment (A) and direct COD measurements (B).

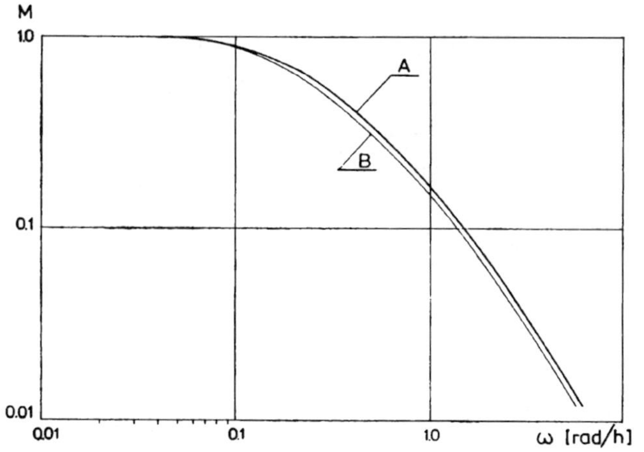

Fig. 8. Comparison of frequency characteristics of models derived from radiotracing examinations (A) and direct COD measurements (B).

a model consisting of two ideal mixing tanks of different time constants.

Some differences between the two models result more from to rare sampling (every 1/2 h) and analytical errors in COD determination than from imperfection of the radiotracing method.

The radiotracing method renders possible to record practically continuously the radiotracer concentration and consequently to obtain more accurate output function.

6. Conclusions

1. The comparison between a model being derived from radiotracing measurements and that being derived from results of process variable change measurements has confirmed high accuracy of the radiotracing method.

2. The radiotracing method, because of continuous measurements, is more accurate one than direct measurement of process variable changes since in the latter case it is not always possible to carry on continuous measurements.

3. In the case when it is necessary to use analytical methods for determining concentrations the radiotracing method is more simple and less expensive one.

Received February 14, 1986

Accepted in revised form March 11, 1986

References

[1] *A. G. Chmielewski, Z. Paca*, Gospodarka Wodna, **46** (1986) 169
[2] *P. Wolf, W. Resnick*, Ind. Eng. Chem., Funds **2** (1963) 287
[3] *K. Imhoff, K. R. Imhoff*, Taschenbuch der Stadtentwässerung, München – Wien: Oldenbourg-Verlag 1979

Isotopenpraxis 22 (1986) 11, S. 411 bis 416

Vergleichende Untersuchungen an ^{85}Kr und ^{133}Xe als Tracer

M. Gläser, W. Klich, A. Creifelds

Akademie der Landwirtschaftswissenschaften der DDR, Forschungszentrum für Mechanisierung der Landwirtschaft, Schlieben/Bornim, DDR-7912 Schlieben, Gartenstraße 30

Auf der Grundlage theoretischer und experimenteller Untersuchungen wurde überprüft, inwieweit ^{85}Kr durch ^{133}Xe als Tracer bei Strömungs-, Volumenstrom-, Gasaustausch- und Vermischungsmessungen ersetzt werden kann. Insbesondere wurde die Effektivität verschiedener Detektortypen und der Empfangsradius in Luft gegenüber der Strahlung dieser Radionuklide ermittelt. Es zeigte sich, daß trotz Einsatz des effektivsten, betaempfindlichen Zählrohrtyps zur Erzielung vergleichbarer Impulsdichten ein höherer Aktivitätseinsatz bei ^{133}Xe im Vergleich zu ^{85}Kr erforderlich ist. Der Einsatz von Szintillationsmeßsonden erhöht die Effektivität, allerdings auf Kosten der räumlichen Auflösung.

On the basis of theoretical and experimental investigations the possibility of substituting the radionuclide ^{85}Kr by the radionuclide ^{133}Xe as tracer for flow-, hold-up time (residual time)-, exchange rate- and mixing measurements in gases was tested. Especially the efficiency of several different counters and the range of the radiation in air (receiving radius) from these radionuclides was estimated. The results show that despite applying the commercially available, most effective β-sensitive counters for achieving similar pulse rates a higher activity of ^{133}Xe in comparison to used activity of ^{85}Kr is needed. Using scintillation measuring probes the effectivity is increased, but at the expense of the spatial resolution.

Keywords

flow rate; gas flow; krypton 85; measuring instruments; mixing; tracer techniques; velocimeters; velocity; xenon 133

1. Einleitung

Für Traceruntersuchungen des Strömungs-, Verweilzeit-, Austausch- und Mischverhaltens von Gasen, insbesondere in technischen Anlagen stehen eine Reihe von Radionukliden für spezielle Einsatzfälle zur Verfügung [1, Tab. 1]. Es muß in jedem Fall eine Auswahl des geeignetsten Radionuklids erfolgen. Besonders verbreiteten Einsatz fand und findet bei Messungen ohne Wanddurchstrahlung das Radionuklid ^{85}Kr (z.B. [1], [2], [3], [4], [13]). Nachteilig ist die Halbwertszeit von 10,76 a, die aus Umweltschutzgründen zur Limitierung der Einsatzaktivität zwingt. Es wird darauf orientiert, Radionuklide mit ähnlichen Eigenschaften wie ^{85}Kr, aber kürzeren Halbwertszeiten zu verwenden. Hier bietet sich ^{133}Xe an, das eine Halbwertszeit von nur 5,25 d besitzt.

Mit den im folgenden dargestellten Untersuchungen sollte überprüft werden, inwieweit ^{85}Kr durch ^{133}Xe insbesondere bei Luftwechselzahl- und Volumenstrommessungen [10] ersetzt werden kann. Dabei interessieren vor allem das jeweilige „*Empfangsvolumen*" (radius), aus dem die Strahlung erfaßt wird, und die *Effektivität* der verfügbaren Detektoren.

2. Theoretische Untersuchungen

Um die Eignung eines Radionuklides als Tracergas einschätzen zu können, ist die Kenntnis des Massenschwächungskoeffizienten der emittierten Kernstrahlung von wesentlicher Bedeutung, da sich damit Aussagen über das Empfangsvolumen des Detektors, über die Meßmöglichkeit durch Trennwände hindurch, über den erforderlichen Detektortyp u. a. treffen lassen.

Bei den zu vergleichenden Radionukliden handelt es sich um Betastrahler (Tab. 1). Im Falle des ^{85}Kr ist die begleitende Gammastrahlung vernachlässigbar, im Falle ^{133}Xe muß die weiche Gammastrahlenkomponente jedoch mit berücksichtigt werden. Deshalb sollen zunächst die Massenschwächungskoeffizienten für beide in Abhängigkeit

Tab. 1. Gegenüberstellung interessierender Eigenschaften von ^{85}Kr und ^{133}Xe (u. a. [11])

Eigenschaft	(Effektive) Kernladungszahl	(Effektive) Massenzahl	^{85}Kr	^{133}Xe
Chemische Natur			Edelgas	Edelgas
Halbwertszeit			10,76 a	5,25 d
Max. Betastrahlungsenergie/keV (und relative Häufigkeit/%)			672(99)	347(100)
Gammastrahlungsenergie /keV (und relative Häufigkeit/%)			513(0,7)	81(100)
Massenschwächungskoeffizienten μ/ϱ				
Betastrahlung/cm² · mg⁻¹				
a) Aluminium	13	27	0,0296	0,0737
b) Glas	10,6	19,7	0,0331	0,0824
c) Glimmer	9,4	18,9	0,0306	0,0762
d) Luft	7,5	14,45	0,0319	0,0794
Gammastrahlung/cm² · g⁻¹				
a) Aluminium			0,084	0,189
b) Glas			0,084	0,168
c) Glimmer			0,085	0,165
d) Luft			0,087	0,162
e) Eisen			0,083	0,550
Preis M/GBq			30	110

von in Frage kommenden Absorbermaterialien berechnet werden.

2.1. *Massenschwächungskoeffizienten für ^{85}Kr- und ^{133}Xe-Strahlung*

Zwischen dem Massenschwächungskoeffizienten μ/ϱ (cm^2 $\times$ g^{-1}) und der maximalen Betastrahlungsenergie E_{max} (in MeV) besteht folgende, für Aluminium gültige, empirische Beziehung [5, S. 465]:

$$\frac{\mu}{\varrho} = \frac{17{,}1}{E_{max}^{1,38}} \qquad (1)$$

Die nach dieser Beziehung berechneten Massenschwächungskoeffizienten (Tab. 1) für ^{85}Kr und ^{133}Xe stimmen mit Literaturangaben gut überein (Abb. 1) [6], [7].

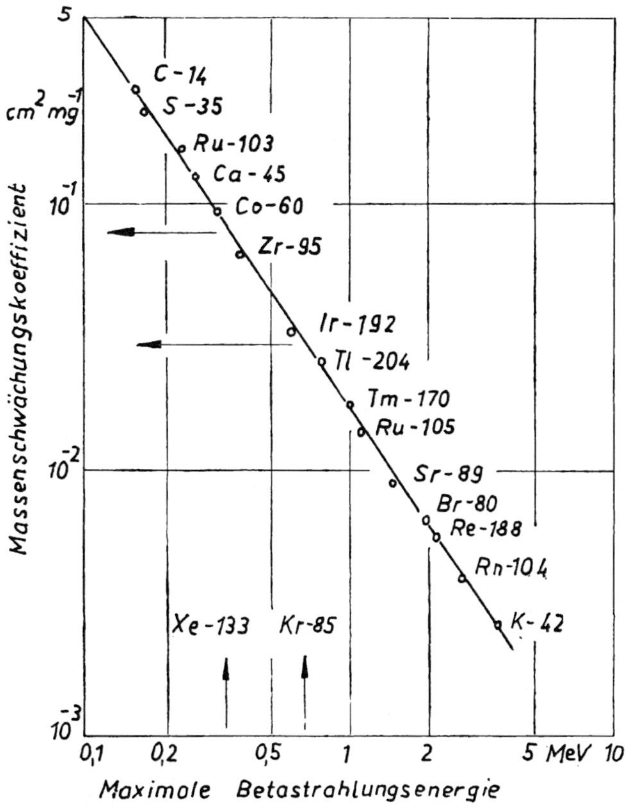

Abb. 1. Massenschwächungskoeffizient in Abhängigkeit von der Maximalenergie der Betastrahlung (nach [6])

Für die Ermittlung der Massenschwächungskoeffizienten anderer Materialien ($Z < 13$) aus den Aluminiumwerten gilt folgende Beziehung [5, S. 465]:

$$\left(\frac{\mu}{\varrho}\right)^{Z} = \left(\frac{\mu}{\varrho}\right)^{Al} \cdot \frac{Z \cdot 27}{A \cdot 13} \qquad (2)$$

Damit erhält man dann auch die μ/ϱ-Werte für die anderen hier interessierenden Materialien (Tab. 1).

Da beim ^{133}Xe-Betazerfall gleichzeitig ein energiearmes Gammaquant emittiert wird, ist mit einem beträchtlichen Anteil von Detektorimpulsen zu rechnen, die von Gammaquanten ausgelöst werden. Wegen ihres größeren Durchdringungsvermögens ergibt sich deshalb ein gegenüber der reinen Betastrahlung vergrößertes Empfangsvolumen. Bei der Berechnung müssen daher beide Anteile berücksichtigt werden.

2.2. *Effektivität*

Im folgenden soll in Anlehnung an *G. Aniansson* u. a. [8] abgeschätzt werden, inwieweit sich die Teilchen- bzw. Photonenrate im empfindlichen Detektorvolumen und damit die Effektivität der Detektoren unter sonst gleichen Bedingungen ändert, wenn das Tracergas ^{85}Kr durch ^{133}Xe ersetzt wird.

Eine radioaktive Gasmenge (Abb. 2) emittiere N Teilchen (oder Photonen) pro Zeiteinheit. Ein Volumenelement sendet dann $N \cdot dV$ Teilchen (oder Photonen) stochastich in den 4π-Raum. Das Oberflächenelement dA des Detektors wird von der differentiellen Teilchen- (oder Photonen)-rate $d\varphi$ gemäß

$$d\varphi = N \cdot dV \cdot \frac{dA \cdot \cos\lambda}{4\pi d_1^2} \qquad (3)$$

durchsetzt. „d_1" ist hierbei die Entfernung des Volumenelementes dV vom Oberflächenelement dA.

Durch die Schwächung der Strahlung im durchstrahlten Medium zwischen jeweiligem Volumenelement und der Oberfläche des Detektors (Abstand d_1) und in der Detektorwandung (Dicke d_2) gelangt nur ein Teil der Teilchen (oder Photonen) in das empfindliche Volumen des Detektors und damit zur Registrierung. Gleichung (3) geht dadurch über in:

$$d\varphi = N \cdot dV \frac{dA \cdot \cos\lambda}{4\pi d_1^2} \cdot e^{-\left[\left(\frac{\mu}{\varrho}\right)_1 \cdot \varrho_1 \cdot d_1 + \left(\frac{\mu}{\varrho}\right)_2 \cdot \varrho_2 \cdot d_2/\cos\lambda\right]} \qquad (4)$$

Nach Einführung von Polarkoordinaten für dV

$$dV = d_1^2 \cdot \sin\lambda \cdot d\delta \cdot d\lambda \cdot dd_1 \qquad (5)$$

und Integration über den Gesamtraum sowie die Detektoroberfläche A erhält man die Gesamtzahl der in das empfindliche Volumen des Detektors eingedrungenen Teilchen oder Photonen pro Zeiteinheit:

$$\varphi = \frac{N}{4\pi} \int_0^{d_1} \int_0^{\pi/2} \int_0^{\pi} \int_0^{A} \sin\lambda \cdot \cos\lambda \times e^{-\left[\left(\frac{\mu}{\varrho}\right)_1 \cdot \varrho_1 \cdot d_1 + \left(\frac{\mu}{\varrho}\right)_2 \cdot \varrho_2 \cdot d_2/\cos\lambda\right]} \cdot dA \cdot d\delta \cdot d\lambda \cdot dd_1 \qquad (6)$$

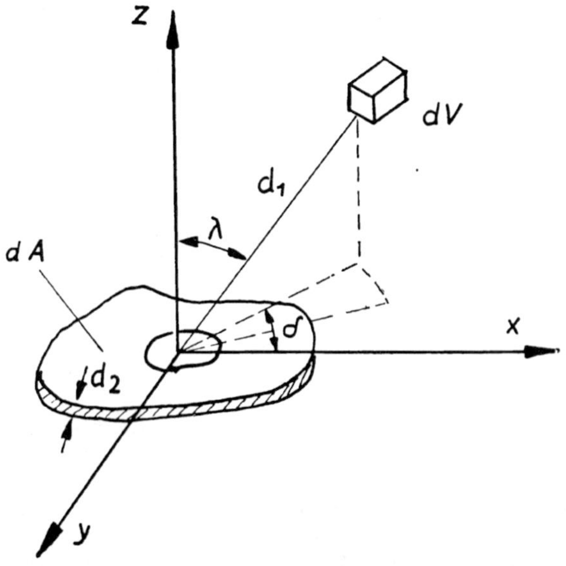

Abb. 2. Zur Ableitung der aus einem Gasraum kommenden und eine Detektorfläche durchsetzenden Teilchen- oder Photonenrate

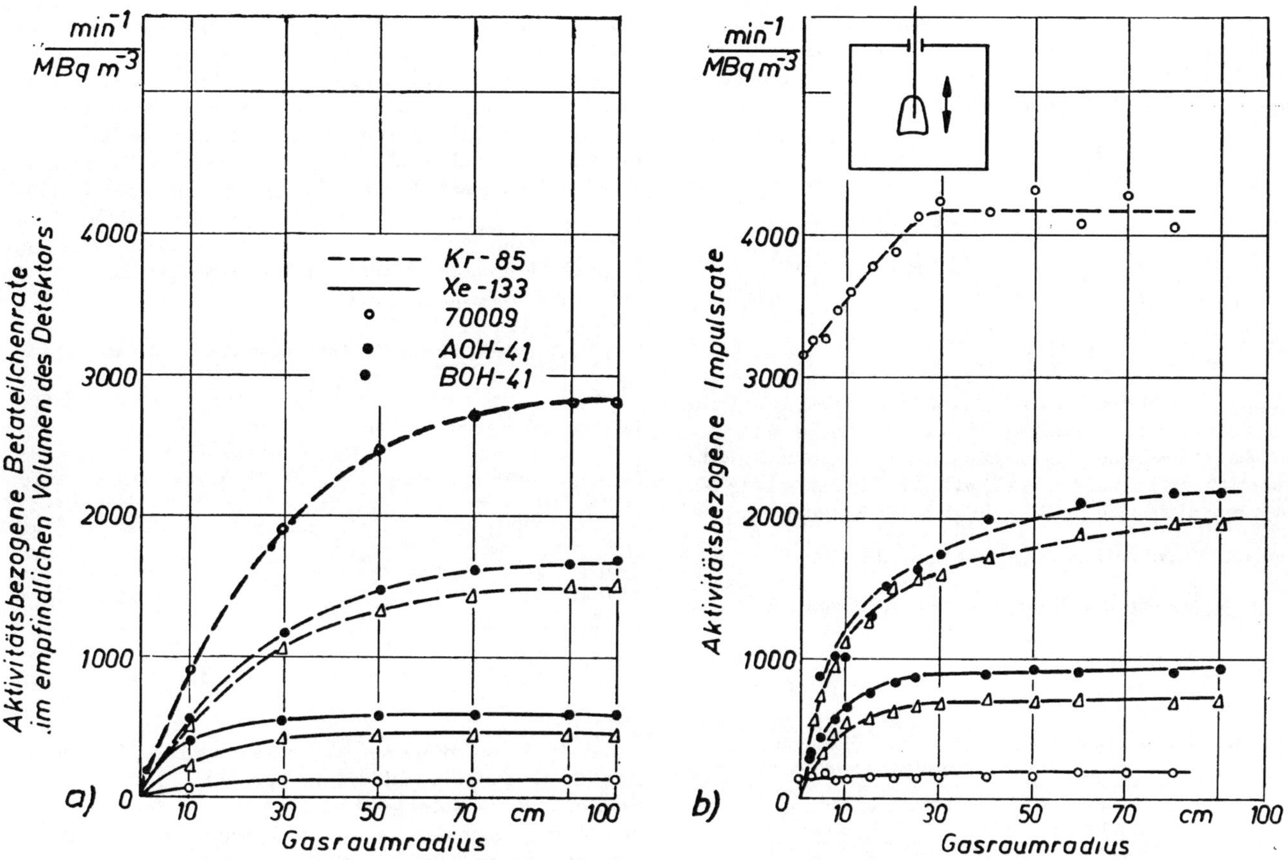

Abb. 3. a) Theoretische, aktivitätsbezogene Betateilchenrate als Funktion des Gasraumradius; b) Experimentelle Ermittlung von Empfangsradius und Effektivität verschiedener Detektortypen gegenüber ^{85}Kr und ^{133}Xe-Strahlung

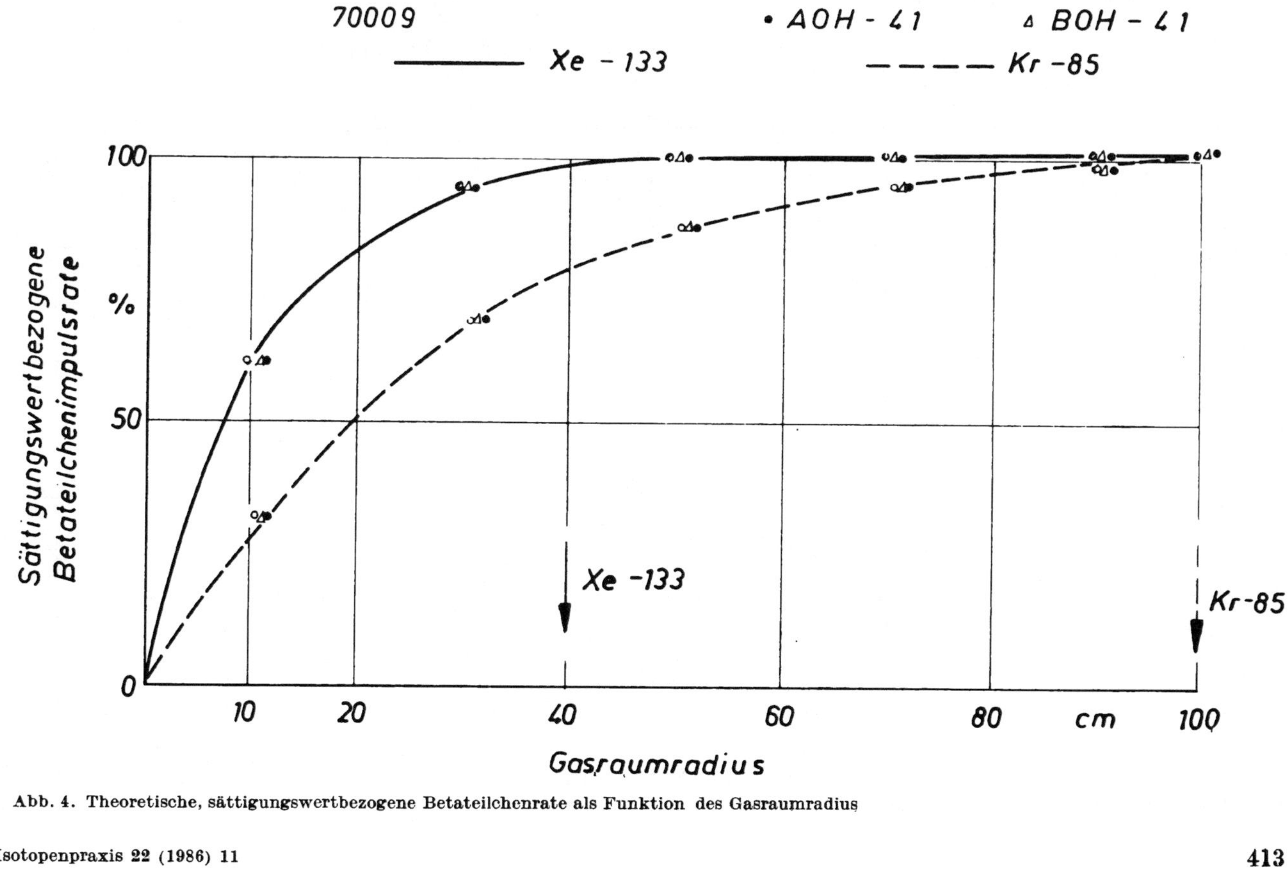

Abb. 4. Theoretische, sättigungswertbezogene Betateilchenrate als Funktion des Gasraumradius

Man erhält unter der Annahme, daß der Radius des betrachteten Gasraumes sehr viel größer ist als der Radius des Zylinderzählrohres bzw. der des Fensters beim Fensterzählrohr, folgende Lösung für den Ansatz (6)

$$\varphi = \frac{A \cdot N}{4} \left\{ \frac{1 - e^{-\left(\frac{\mu}{\varrho}\right)_1 \varrho_1 d_1}}{\left(\frac{\mu}{\varrho}\right)_1 \cdot d_1} \right\} \times \left\{ \left[1 - \left(\frac{\mu}{\varrho}\right)_2 \cdot \varrho_2 \cdot d_2\right] e^{-\left(\frac{\mu}{\varrho}\right)_2 \varrho_2 d_2} - \left[\left(\frac{\mu}{\varrho}\right)_2 \varrho_2 d_2\right]^2 \times Ei\left[-\left(\frac{\mu}{\varrho}\right)_2 \varrho_2 d_2\right] \right\} \tag{7}$$

Hierbei berücksichtigt der 1. Term die spezifische Aktivität im Gasraum und die empfindliche Detektorfläche, der 2. Term die Strahlenschwächung im Gas und der 3. Term die Strahlenschwächung im Fenster des Detektors. Die numerische Auswertung dieses Ausdrucks erfolgt unter Nutzung der tabellierten Exponentialintegralfunktion [9] für $Ei\left[-\left(\frac{\mu}{\varrho}\right)_2 \times \varrho_2 \cdot d_2\right]$ sowie der Kenndaten der in Frage kommenden Detektoren (Tab. 2).

Tab. 2. Kenndaten der getesteten Detektoren

Detektor	Empfindliche Fläche/cm²	Fenster- bzw. Wandmaterial	Flächenmasse $\varrho_2 \cdot d_2$/mg · cm^{-2}
AOH-41	4,9	Glimmer	2
BOH-41	4,9	Glimmer	4
70009(VA-Z-118)	45,0	Glas	35 ... 55
70013(VA-Z-221)	70	Glas, Edelstahl	≈ 400
VA-S-963	260	Aluminium	≈ 500
27000 (VA-S-968) mit SKG 1 S 1	260	Aluminium	≈ 500
72016 (VA-S-50) mit SKG 1 SN 6	25	Aluminium	≈ 500

Die nach Gleichung (7) berechneten Betaraten sind für ^{85}Kr- bzw. ^{133}Xe-Markierung und drei betaempfindliche Detektortypen in Abhängigkeit vom Gasraumradius dargestellt (Abb. 3a, 4). In allen Fällen steigt zunächst erwartungsgemäß die in den Detektor gelangende Teilchenrate an, mit zunehmender Größe des Gasraumes nähert sie sich bei der Betastrahlung relativ rasch einem Sättigungswert (Abb. 3a und 4, Tab. 3). Dieser wird jeweils bei 95% des „unendlichen" Wertes entnommen. Der jeweilige Beginn dieses Sättigungsbereiches wird als Empfangsvolumen (-radius) bezeichnet.

Tab. 3. Die Empfangsradien für Betastrahlung von ^{85}Kr und ^{133}Xe in Luft

Detektor	Empfangsradien/cm			
	theoret.	exper.	theor.	exper.
	^{85}Kr		^{133}Xe	
AOH-41	100	90	40	30
BOH-41	100	90	40	30
70009	100 (Abb. 3a)	100 (Abb. 3b u. 7)	40 (Abb. 3a)	60 (Abb. 3b u. 7)

Bei der Berechnung der aktivitätsbezogenen Photonenrate für die Gammastrahlungskomponente des ^{133}Xe zeigte sich, daß die Strahlungsschwächung in den Fenstern bzw. den Wandungen der getesteten Detektoren zu vernachlässigen ist, und daß ein Sättigungswert erst weit über 10 m zu erwarten ist (Abb. 5). Das bedeutet, daß die Detektoren von einer relativ intensiven Photonenflußdichte durchsetzt werden und damit auch ein hoher Anteil der Gesamtimpulsraten auf die Gammastrahlung zurückzuführen und ein größerer Empfangsradius als der nur für den Betastrahlenanteil berechnete Wert zu erwarten ist (vgl. Abb. 7).

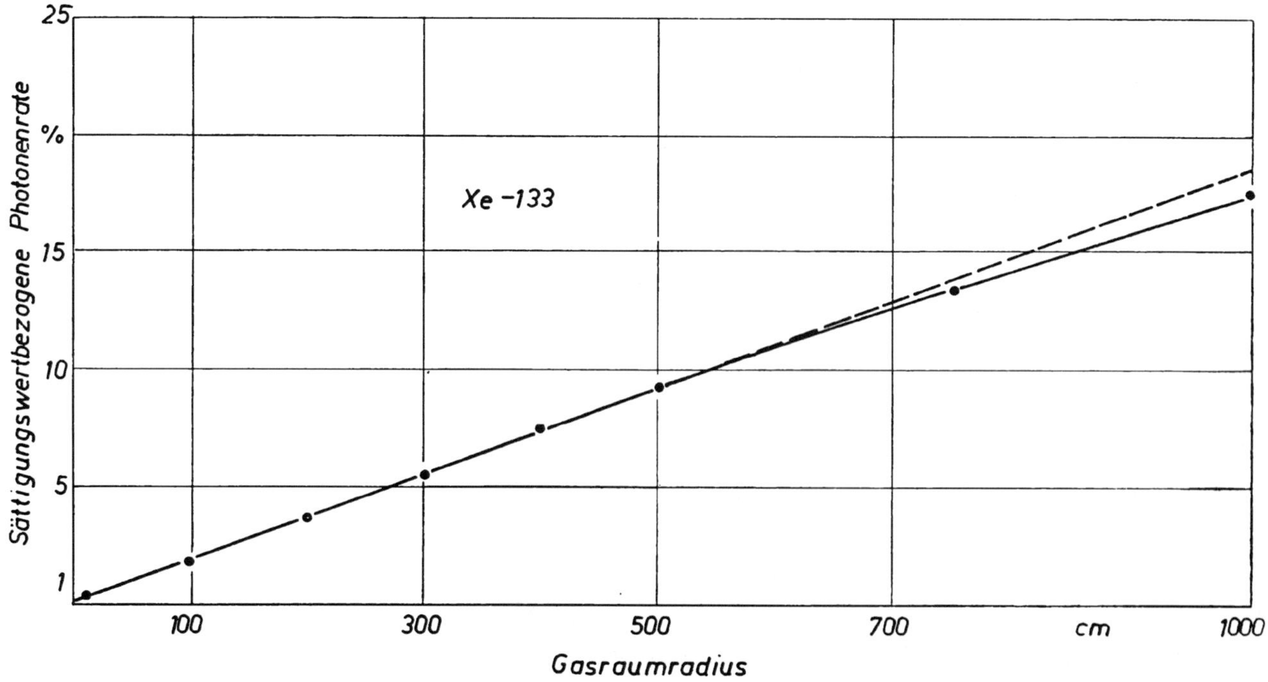

Abb. 5. Theoretische, aktivitätsbezogene Photonenrate als Funktion des Gasraumradius

3. Experimentelle Bestimmung von Effektivität und Empfangsvolumen

Zur experimentellen Ermittlung von Effektivität und Empfangsvolumen der untersuchten Detektoren wurden Messungen in einem 1 m³ großen, gasdichten, würfelförmigen Experimentierbehälter durchgeführt (Abb. 6). In diesen Raum wurde eine ^{85}Kr- bzw. ^{133}Xe-Gasmenge jeweils möglichst gleicher Aktivität eingebracht und gleichmäßig verteilt.

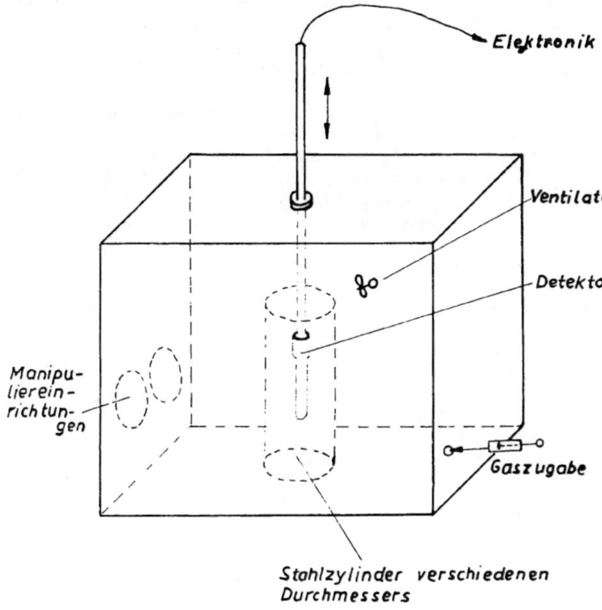

Abb. 6. Experimentierbehälter

3.1. *Empfangsvolumen*

Zur Ermittlung des Empfangsvolumens (-radius) von Fensterzählrohren wurden die von außen manipulierbaren Detektoren in verschiedene vertikale Abstände vom Boxboden gebracht und so deren „Empfangsvolumen (-radius)" für die aus dem Gasraum kommende ionisierende Strahlung variiert. Demnach liegen die experimentell ermittelten Empfangsradien (Abb. 3b) dieser Fensterzählrohre in der Nähe der theoretisch berechneten Werte (Abb. 3a, Tab. 3).

Bei den getesteten zylinderförmigen Zählrohren diente diese Versuchsvariante lediglich der Ermittlung der Höhe der Sättigungswerte. Die (seitlichen) Empfangsradien wurden mittels folgender Versuchsvariante erhalten: Die etwa in der Mitte des Experimentierbehälters befindlichen Zählrohre wurden von Eisenhohlzylindern unterschiedlichen Durchmessers und einer Wanddicke von 10 mm zentrisch umgeben, und die Impulsrate als Funktion des Radius dieser Zylinder aufgenommen (Abb. 7, Tab. 3).

3.2. *Effektivität*

Zur Ermittlung der Effektivität der verfügbaren Detektoren gegenüber ^{85}Kr- und ^{133}Xe-Strahlung wurden diese unter gleichen Bedingungen sowohl in dem angegebenen Experimentierbehälter (Abb. 6) als auch in einem größeren Raum miteinander verglichen, wobei die nur auf Gammastrahlung ansprechenden Detektoren wegen des gegen-

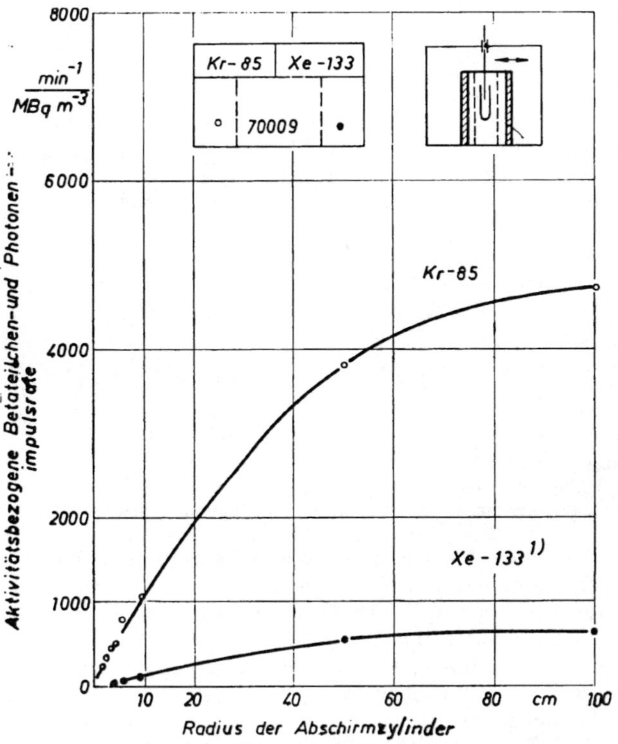

Abb. 7. Zur Ermittlung des Empfangsvolumens von zylinderförmigen Zählrohren. [1]) Impulsrate verdreifacht dargestellt

über den Betastrahlen wesentlich größeren Empfangsvolumens (Abb. 5) sinnvollerweise lediglich beim Raumversuch mit ^{133}Xe mit eingesetzt wurden.

4. Ergebnisse

Die Vergleichsuntersuchungen an ^{85}Kr und ^{133}Xe bezüglich ihrer Eignung als radioaktiver Tracer für Gase ergaben, daß bei ^{133}Xe der Empfangsradius bei Einsatz von im wesentlichen betaempfindlichen Detektoren nur etwa ein Viertel desjenigen von ^{85}Kr beträgt, das Empfangsvolumen sich demzufolge auf etwa 1/15 reduziert. Das führt zu einer stark verringerten Effektivität der untersuchten Detektortypen beim ^{133}Xe-Einsatz. Bedingt durch die begleitende Gammastrahlung beim ^{133}Xe ist der Effektivitätsrückgang jedoch nicht so groß wie der geometrischen Abschätzung unter Vernachlässigung des Gammaanteiles entspräche. Bei den Fensterzählrohren vom Typ AOH-41 und BOH-41 kommt außerdem die wesentlich geringere Betastrahlungsschwächung im Fenster gegenüber den Zylinderzählrohren begünstigend hinzu. Das führt schließlich zu dem Ergebnis, daß, obwohl die Fensterzählrohre die Betastrahlung nur aus dem Halbraum empfangen, diesen Detektortypen, insbesondere dem Typ AOH-41, beim Übergang von der ^{85}Kr- zur ^{133}Xe-Tracertechnik der Vorrang vor den anderen kommerziell verfügbaren betaempfindlichen Detektoren zu geben ist (Abb. 3b und 7, Tab. 4).

Anders sieht die Situation beim Einsatz von gammaempfindlichen Detektoren aus. Hier ist — bedingt durch das um Größenordnungen ausgedehntere Empfangsvolumen — eine wesentliche Effektivitätserhöhung durch Erfassung der weichen Gammastrahlung des ^{133}Xe gegeben (Abb. 5 und Tab. 4), allerdings auf Kosten der räumlichen Auflösung der Meßmethode.

Tab. 4. Relative Effektivität verschiedener Detektoren gegenüber ^{85}Kr und ^{133}Xe-Strahlung

Detektor	^{85}Kr			^{133}Xe		
	*)	**)	theor.	*)	**)	theor.
AOH-41	0,53	0,60	0,71	0,21	0,18	0,26
BOH-41	0,47	0,59	0,66	0,16	0,16	0,22
70009***) (VA-Z-118)	1,00	1,00	1,00	0,04	0,10	0,10
70013 (VA-Z-221)	–	–	–		0,06	–
VA-S-963	–	–	–	–	23	–
27000 (VA-S-968) mit SKG 1-S1	–	–	–	–	24	–
72016 (VA-S-50) mit SGK 1 SN 6	–	–	–	–	11	–

*) Box-Versuch
**) Raum-Versuch
***) Effektivität = 1 gesetzt bei ^{85}Kr und Zählrohr 70009

Auf Grund dieser Ergebnisse und der gültigen Preise für die Tracergase kann in den Fällen, wo bei der Untersuchung keine hohe räumliche Auflösung gefordert wird, und die Versuche zeitlich einigermaßen planbar sind, ^{133}Xe, durchaus ökonomisch eingesetzt werden. Allerdings wird auf den Einsatz von ^{85}Kr nicht ganz verzichtet werden können, insbesondere bei Langzeituntersuchungen und umfangreicheren Untersuchungen mit hoher räumlicher Auflösung; im letzteren Fall müßten zur Erzielung vergleichbarer Impulsraten wesentlich höhere Aktivitäten eingesetzt werden als beim ^{85}Kr.

Eingegangen am 11. 2. 1986

Literatur

[1] *R. Otto, P. Hecht, H. Ulrich, H. Häntzschel,* Isotopenpraxis **14** (1978) 120
[2] *M. Beer, A. Heißner, G. Vogel,* Deutsche Agrartechnik **15** (1965) 1666
[3] *M. Gläser, H.-J. Müller,* Wissenschaft und Fortschritt **33** (1983) 350
[4] *K. Irmer,* Isotopenpraxis **12** (1976) 333
[5] *G. Brunner,* Verfahren zur Untersuchung von Stoffzusammensetzungen durch Messung von Streuung und Absorption von Kernstrahlung und Anregung von Sekundärstrahlung, in: *W. Hartmann* (Hrsg.), Meßverfahren unter Anwendung ionisierender Strahlung. Leipzig: Akademische Verlagsgesellschaft Geest und Portig, K.-G. 1969
[6] *L. Meyer-Schützmeister, D. Vincent, H. Houtermanns,* S. 348–360, in: *Landolt-Börnstein,* Zahlenwerte und Funktionen, Bd. 1, Teil 5: Atomkerne und Elementarteilchen (6. Auflage, Berlin – Göttingen – Heidelberg 1952)
[7] *G. J. Gleason,* Nucleonics 8(5) (1951/12)
[8] *G. Aniansson, G. Anniansson, O. Noreen,* Technik Tidskrift, Stockholm 10. Jan. 1956, 17
[9] *Jahnke-Emde,* Tafeln höherer Funktion. Leipzig: B. G. Teubner 1960
[10] *A. Creifelds, W. Klich, M. Gläser,* Verfahren und Einrichtung zur Volumenstrommessung mittels radioaktiver Tracer, Patenthinterlegung (Reg. Nr. 601 F/2 68492/6)
[11] *R. Jaeger* u. a. (Hrsg.), Engineering Compendium on Radiation Shielding, Vol. I. Berlin, Heidelberg, New York: Springer-Verlag 1968

Isotopenpraxis 22 (1986) 11, S. 416

Buchbesprechung

The Handbook of Environmental Chemistry

Vol. 4, Part A: Air Pollution. O. Hutzinger (ed.). Berlin – Heidelberg – New York – Tokyo: Springer Verlag 1986, 222 S., 71 Abb., 21 Tab.: geb. DM 168,–

Thomas Mann schreibt in seinem „Zauberberg", daß Zeit Lethe sei, „aber auch Fernluft ist so ein Trank, und sollte sie weniger gründlich wirken, so tut sie es dafür desto rascher." Denkt man heute nicht bei dem Wort Fernluft an Langstreckenverfrachtungen von Luftverschmutzungen, an die Luftverschmutzung schlechthin? Im Vorwort zu dem vorliegenden ersten Teilband heißt es, daß sich der Herausgeber entschlossen hat, den Problemen der Luftverschmutzung einen gesonderten Band, den 4. Band des Handbuches, ihrer besonderen Stellung in der Umweltchemie entsprechend, einzuräumen. Natürlich war man sich der Schwierigkeit einer scharfen Grenzziehung zu den anderen umweltchemischen Aspekten bewußt, hängt doch die Luftverschmutzung auch und nicht unwesentlich mit den biogeochemischen Kreisläufen (CO_2-Problem beispielsweise) oder mit Reaktionen und Vorgängen (der Luftchemie) zusammen, wie sie auch in den anderen Bänden bereits behandelt wurden.

Vorangestellt wird diesem Teilband ein kurzer Abriß der Luftverschmutzung, der schließlich in einigen perspektivischen Betrachtungen endet. Dieser Abschnitt ist von einem englischen Wissenschaftler verfaßt – *A. Witt,* University of Nottingham –, und so ist auch der historische Teil auf England bezogen. Bereits im Jahre 1273 wurde die Verbrennung von Kohle in London verboten, sie sei schädlich für die Gesundheit. Dieses Datum ist insofern bemerkenswert, als der englische König *Eduard I.* noch nicht gekrönt war und er außerdem noch gar nicht von seinem Kreuzzug ins Mutterland zurückgekehrt war. Kurz vor seinem Tode proklamierte der gleiche König, daß Steinkohle in Öfen nicht verfeuert werden dürfe. Aber bereits im Jahre 1240 hat sich der Stauferkaiser *Friedrich II.* in einem Edikt über die Reinhaltung der Luft, der Gewässer und des Bodens ausgesprochen: der kontinentale Teil des historischen Abrisses.

Die anderen Kapitel des Buches sind den zu erwartenden Luftverschmutzungen gewidmet: Halogenkohlenwasserstoffe, photochemischer Smog. Zwei weitere Kapitel haben die Modellierungen der luftchemischen Prozesse zum Gegenstand, also einerseits die Modelle der Dispersion und andererseits die Vorstellungen über den Austrag von Luftverschmutzungen aus der Atmosphäre. Dabei werden Vor- und Nachteile einzelner Modellansätze – man möchte sagen – fast wertfrei nebeneinander gestellt, um dem Leser Einblicke in die Vielfalt solcher Vorstellungen zu ermöglichen; ein besonders hervorzuhebender Gesichtspunkt, der dem Anspruch, ein Handbuch zu sein, sehr gerecht wird.

Auch dieser Teilbestand bietet dem Wissenssuchenden eine reiche Auswahl modernster Literatur: insgesamt sind 595 Literaturstellen angegeben, allein aus den Jahren 1980–1985 290 Zitate ($\triangleq$ 48,7%). Daß dieser Band weitere Teilbände erwarten läßt, liegt sehr nahe; denn an Themen darüber mangelt es nicht (Chemie der Rauchfahnen, Waldschäden und Luftverschmutzung, Aerosole, Stickoxide als Umweltfaktoren u. ä. m.). Ein Leserkreis für den vorliegenden Teilband wird sich schnell einfinden; denn einige grundlegende Tatsachen sind mit hohem Sachverstand und in gelungener Auswahl bereits dargestellt.

H. Hübner

ISOTOPENPRAXIS

Band 22 1986 Heft 11

Contents / Inhalt

Authors	Page	Articles	Originalarbeiten
J. Guizerix	377	▶Trends and Open Questions in Industrial Tracer Applications	Trends und offene Fragen der industriellen Traceranwendung
A. M. Eid, F. I. A. Said, S. M. Kamal	381	▶Ceric-cerous Sulphate and TLD Techniques for Radiation Processing Control in Egypt	Ce(IV)/Ce(III)-Sulfate und die Thermolumineszenz-Dosimetrie (TLD) für die strahlentechnische Kontrolle in Ägypten
J. Rant, R. Ilić, G. Pregl, S. Demirović	385	▶Neutron Radiographic Detection of Surface Corrosion on Aluminium	Neutronenradiographischer Nachweis der Oberflächenkorrosion von Aluminium
F.-D. Kopinke, J. Dermietzel, W. Jockisch, G. Räuber	388	Determination of the Isotope Distribution in ^{14}C Labelled Hydrocarbons Through Analysis of the Pyrolysis Products	▶Bestimmung der Isotopenverteilung in ^{14}C-markierten Kohlenwasserstoffen durch thermische Fragmentierung
J. Hons	392	Determination of the Depth of the Liquid Phase in Continuous Casting	▶Die Bestimmung der Tiefe der flüssigen Phase beim Stranggruß
N. Stollárová, L. Mráz	395	Kinetics of the Distribution of Iodine-131 After an Acute Pansomatic Exposure With a Radiation of 390 mC · kg^{-1}	▶Die Kinetik der Verteilung von ^{131}I nach einer akuten pansomatischen Exposition von 390 mC · kg^{-1}
E. Winkler, H. Sprinz, B. Lindemann, U. Hädrich, U. Franck	400	Computer-based Processing and Interpretation of NMR Spectra in the Field of Isotope and Radiation Research	▶Rechnergestützte Verarbeitung und Auswertung von NMR-Spektren in der Isotopen- und Strahlenforschung
A. G. Chmielewski, A. Dobrowolski	407	▶Verification of the Usefulness of Radiotracing Methods in Hydrodynamic Investigations	Überprüfung der Nützlichkeit von Tracermethoden für hydrodynamische Untersuchungen
M. Gläser, W. Klich, A. Creifelds	411	Comparative Investigations of ^{85}Kr and ^{133}Xe as Tracers	Vergleichende Untersuchungen an ^{85}Kr und ^{133}Xe als Tracer
	416	**New Books**	**Buchbesprechungen**

▶Language of Publication / ▶Publikationssprache

Isotopenpraxis is indexed in Current Contents/PC & ES, Science Citation Index, and INIS Atomindex